D1000048

Functional Analysis
in Applied
Mathematics
and Engineering

Studies in Advanced Mathematics

Series Editor

STEVEN G. KRANTZ
Washington University in St. Louis

Editorial Board

R. Michael Beals
Rutgers University

Dennis de Turck
University of Pennsylvania

Ronald DeVore
University of South Carolina

Lawrence C. Evans
University of California at Berkeley

Gerald B. Folland
University of Washington

William Helton
University of California at San Diego

Norberto Salinas
University of Kansas

Michael E. Taylor
University of North Carolina

Titles Included in the Series

Steven R. Bell, The Cauchy Transform, Potential Theory, and Conformal Mapping

John J. Benedetto, Harmonic Analysis and Applications

John J. Benedetto and Michael W. Frazier, Wavelets: Mathematics and Applications

Albert Boggess, CR Manifolds and the Tangential Cauchy–Riemann Complex

Goong Chen and Jianxin Zhou, Vibration and Damping in Distributed Systems, Vol. 1: Analysis, Estimation, Attenuation, and Design. Vol. 2: WKB and Wave Methods, Visualization, and Experimentation

Carl C. Cowen and Barbara D. MacCluer, Composition Operators on Spaces of Analytic Functions

John P. D'Angelo, Several Complex Variables and the Geometry of Real Hypersurfaces

Lawrence C. Evans and Ronald F. Gariepy, Measure Theory and Fine Properties of Functions

Gerald B. Folland, A Course in Abstract Harmonic Analysis

José García-Cuerva, Eugenio Hernández, Fernando Soria, and José-Luis Torrea, Fourier Analysis and Partial Differential Equations

Peter B. Gilkey, Invariance Theory, the Heat Equation, and the Atiyah-Singer Index Theorem, 2nd Edition

Alfred Gray, Modern Differential Geometry of Curves and Surfaces with Mathematica, 2nd Edition

Eugenio Hernández and Guido Weiss, A First Course on Wavelets

Steven G. Krantz, Partial Differential Equations and Complex Analysis

Steven G. Krantz, Real Analysis and Foundations

Kenneth L. Kuttler, Modern Analysis

Clark Robinson, Dynamical Systems: Stability, Symbolic Dynamics, and Chaos, 2nd Edition

John Ryan, Clifford Algebras in Analysis and Related Topics

Xavier Saint Raymond, Elementary Introduction to the Theory of Pseudodifferential Operators

Robert Strichartz, A Guide to Distribution Theory and Fourier Transforms

André Unterberger and Harald Upmeier, Pseudodifferential Analysis on Symmetric Cones

James S. Walker, Fast Fourier Transforms, 2nd Edition

James S. Walker, Primer on Wavelets and their Scientific Applications

Gilbert G. Walter, Wavelets and Other Orthogonal Systems with Applications

Kehe Zhu, An Introduction to Operator Algebras

Functional Analysis in Applied Mathematics and Engineering

MICHAEL PEDERSEN

Released from
Samford University Library

Samford University Library

CHAPMAN & HALL/CRC

Boca Raton   London   New York   Washington, D.C.

Library of Congress Cataloging-in-Publication Data

Pedersen, Michael.
    Functional analysis in applied mathematics and engineering /
Michael Pedersen.
       p. cm. — (Studies in advanced mathematics)
    Includes bibliographical references and index.
    ISBN 0-8493-7169-4 (alk. paper)
    1. Functional analysis. I. Title II. Series.
QA320.P394 1999
515′.7—dc21   99-37641
                                                          CIP

This book contains information obtained from authentic and highly regarded sources. Reprinted material is quoted with permission, and sources are indicated. A wide variety of references are listed. Reasonable efforts have been made to publish reliable data and information, but the author and the publisher cannot assume responsibility for the validity of all materials or for the consequences of their use.

Neither this book nor any part may be reproduced or transmitted in any form or by any means, electronic or mechanical, including photocopying, microfilming, and recording, or by any information storage or retrieval system, without prior permission in writing from the publisher.

The consent of CRC Press LLC does not extend to copying for general distribution, for promotion, for creating new works, or for resale. Specific permission must be obtained in writing from CRC Press LLC for such copying.

Direct all inquiries to CRC Press LLC, 2000 N.W. Corporate Blvd., Boca Raton, Florida 33431.

Trademark Notice: Product or corporate names may be trademarks or registered trademarks, and are only used for identification and explanation, without intent to infringe.

© 2000 by CRC Press LLC

No claim to original U.S. Government works
International Standard Book Number 0-8493-7169-4
Library of Congress Card Number 99-37641
Printed in the United States of America 1 2 3 4 5 6 7 8 9 0
Printed on acid-free paper

QA
320
.P394
2000

Acknowledgments

I would like to thank my colleagues, Prof.O.Jørsboe and Prof.L.Mejlbro, for correcting a number of errors and for collecting so many good exercises. Also, a warm thanks to my former students, Dr.Roger Krishnaswamy, and Dr.Michael Danielsen for the material on the Hilbert Uniqueness Method, and Dr.Anda Binzer for TEX assistance.

Michael Pedersen

Preface

Functional analysis is a "supermodel" for mathematical analysis that has been developed in particular after the Second World War. The general idea is to construct an abstract framework suited to deal with various problems from mathematical analysis that perhaps at first glance seem to have nothing in common, but when all "inessentials" are stripped away, appear to be similar. The language from linear algebra is used to describe the setting of the problems, and this notational simplification is both the power and danger of functional analysis: many formulations and formulas look extremely simple but typically reflect highly delicate subjects.

The present book, *Functional Analysis in Applied Mathematics and Engineering,* is primarily written to cover the course in functional analysis given at The Technical University of Denmark. This is a one-semester (14 weeks) course, and in order to pass, the students will have to do 12 obligatory sets of exercises and have 10 of them judged "passed". Therefore, there is a large number of exercises, many of them are folklore in the sense that they apparently can be found in all basic functional analysis courses. Solutions to all the exercises can be obtained by contacting the author by e-mail. The course is recommended for students on third year or later having a solid background in basic mathematical analysis or corresponding mathematical "maturity".

The aim of the book is to get as fast as possible to the operators on Hilbert spaces with the spectral theorem as the final highlight, without compromising on the mathematical rigor. The course will provide a solid background for the understanding of the problems encountered in applied mathematics and engineering, focusing in particular on the abstract formulations of partial differential equations in a Hilbert space setting. A lot of sophisticated theory that should be incorporated in a pure math. functional analysis course is left out, such as measure theory and more advanced topological considerations. Instead, we typically argue by a density or completion statement. We consider only separable Hilbert spaces since nonseparable Hilbert spaces occur very rarely in applications.

Chapters 1-6 are mandatory; they provide the theory that is necessary in order to read and understand texts and papers in modern applied mathematics. The basic idea of these chapters is to "teach" the subject, and we emphasize that in order to fully understand the theory, it is vital to calculate a large number of the exercises. The exercises are ordered roughly following the chapters, some of them are easy applications of the theory, others are more interesting and it is highly recommended to do at least 50-60 of them. Therefore, it is also recommended that some of the time of the regular lectures is used to discuss the aspects of the theory developed in the exercises. The last chapters are introductions to different subjects from functional analysis and can be read independently. The main purpose of these chapters is to serve as an appetizer to a more specialized study of functional analysis and control theory for partial differential equations. There are no exercises to these chapters, and since the purpose is now completely different from the first part of the book and the theory rather advanced, a number of proofs are omitted and references are much more common. In Chapters 12 and 13, however, detailed proofs of recent results that so far only are published in scientific journals are incorporated. The idea is to take one or two subjects from these chapters as examples of advanced applications as the conclusion of the regular 14-week course, but it is also possible to use the material in a more advanced applied functional analysis course.

There is a vast literature on functional analysis, the main part covering what could be called *general functional analysis* in the sense that the aim is to state and prove the theory in maximal generality, and the reader is typically required to have a good knowledge of measure theory and general topology. This kind of litterature is well suited for pure mathematics students, where the requirements are essential also for many other subjects during the period of study. For an engineering student who must be educated in a number of other technical disciplines, it will typically not be possible to meet the requirements of general functional analysis at the time in the study where the mathematical problems encountered in the technical diciplines become so complicated that they actually require functional analytic considerations. With the present book, we try to meet the background of typical engineering or applied mathematics students and give a short but firm introduction to the subject. As mentioned above, we do not compromise on the mathematical rigor and the "construction" of the book is classical, with theorems and detailed proofs. The main part of the book is focused on operators on Hilbert spaces, and closed unbounded operators are introduced to provide the proper setting for the differential equations from applied mathematics and engineering, some of which are presented in the last part of the book.

Contents

Chapter 1

Topological and Metric Spaces

In this chapter we introduce the most fundamental spaces in functional analysis, the *topological spaces* and the *metric spaces*. Since we will proceed almost immediately to the metric spaces, the topologcal spaces are defined for reference purposes only.

1.1 Some Topology

Perhaps one of the most basic features in analysis is the concept of open and closed sets. More precisely, we define a *topological space* S in the following way: S is a set in which we have a collection of subsets τ satisfying

$$S \in \tau, \quad \emptyset \in \tau \tag{1.1}$$

$$\text{if} \quad A_1, A_2, \ldots A_k \in \tau, \quad \text{then} \quad \cap_1^k A_j \in \tau \tag{1.2}$$

$$\text{if} \quad A_j \in \tau \text{ for all } j \in I, \text{ then } \cup_{j \in I} A_j \in \tau, \tag{1.3}$$

where I is a (not necessarily finite) index set. The sets in τ are denoted *open* sets, and a set is defined to be *closed* if its complement is open. Note that sets can be both closed and open. The *closure* $\overline{A}$ of a set A is the smallest closed set containing A, and the *interior* of a set is the largest open set contained in it. A *neighborhood* of a point x in S is an open set containing x. We call (S, τ) a *Hausdorff Space* and τ a *Hausdorff topology* if every two distinct points in S have disjoint neighborhoods; that is, the topology separates points in S. We define continuity in the following way:

DEFINITION 1.1   *Let (S, τ) and (T, σ) be topological spaces, and let $f : S \to T$ be a mapping. Then f is said to be continuous if $f^{-1}(A)$ is open in S for all open sets A in T.*

If f^{-1} is itself a mapping and also continuous, f is called a *homeomorphism* . (We use the notation $f^{-1}(A)$ for the set $\{x \in S | f(x) \in A\}$, even when f is not assumed to be injective.)

Example 1.1

Let τ in the definition above be *all* subsets of S. Then *all* maps from S to T are continuous.   ⬜

1.2   Metric Spaces

The study of continuous mappings between topological spaces is denoted *general topology*, and is a rich and vital area of mathematics. We will, however, in this context usually demand that the spaces we work in have even more structure. We would like the space to have a natural way to measure distances between elements, giving rise to the notion of *metric spaces.*

DEFINITION 1.2   *Let M be a set and d : $M \times M \to [0; \infty[$ a function. Then (M, d) is denoted a metric space and d a metric if for all x, y, z in M we have:*

$$(i)\quad 0 \le d(x, y)$$
$$(ii)\quad d(x, y) = 0 \quad \text{if and only if} \quad x = y$$
$$(iii)\quad d(x, y) = d(y, x)$$
$$(iv)\quad d(x, y) \le d(x, z) + d(z, y).$$

The condition (iv) above is denoted the triangle inequality.

Example 1.2

The set of real numbers R is a metric space when equipped with the metric $d(x, y) = |x - y|$. This metric is called the *natural* metric on R. Another metric, d', on R is defined from this in the following way:

$$d'(x, y) = \frac{d(x, y)}{1 + d(x, y)}.$$

(See Exercise 6).   ⬜

Notice that any metric space (M,d) is a topological space if equipped with the topology where the topology is stemming from the open balls $B(x_0, r) = \{x \in M \mid d(x_0, x) < r\}$. A set $A \subset M$ is defined to be open if any point in A is in an open ball contained in A. (See Exercise 2.)

Example 1.3

If $M = C([a; b])$, the continuous functions on the finite interval $[a; b]$, we have several interesting metrics:

$$d_\infty(f, g) = \sup_{t \in [a;b]} |f(t) - g(t)| \tag{1.4}$$

$$d_1(f, g) = \int_a^b |f(t) - g(t)| \, dt \tag{1.5}$$

or in general,

$$d_p(f, g) = \left(\int_a^b |f(t) - g(t)|^p \, dt \right)^{\frac{1}{p}}, \qquad \text{for } p \geq 1. \tag{1.6}$$

It is not at all evident that d_p is a metric - but we will see later that these metrics in fact are induced by *norms*. ☐

DEFINITION 1.3   *We say that a sequence* $(x_n) \subset M$ *converges to* $x \in M$ *if* $d(x_n, x) \to 0$ *for* $n \to \infty$.

In this case we write $x_n \to x$ *or* $\lim_n x_n = x$. *The element* $x \in M$ *is denoted the* limit point *for the sequence* (x_n).

Example 1.4

Consider in $C([-1; 1])$ the sequence (f_n) of functions below :

We see that

$$d_1(f_n, 0) = \int_{-1}^1 f_n(t) \, dt = \frac{1}{n} \to 0$$

so $f_n \to 0$ in $(C([-1; 1]), d_1)$. On the other hand, it is obvious that in $(C([-1; 1]), d_\infty)$ we have

$$d_\infty(f_n, 0) = 1 \qquad \text{for all } n,$$

so (f_n) does not converge to 0 in this space. ☐

The triangle inequality ensures that any metric is continuous.

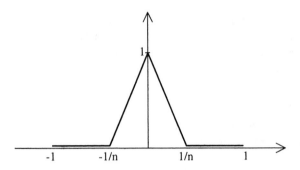

FIGURE 1.1

Graph of (f_n)

PROPOSITION 1.1

If $x_n \to x$ and $y_n \to y$ in (M, d), then $d(x_n, y_n) \to d(x, y)$.

PROOF

$$d(x, y) \leq d(x, x_n) + d(x_n, y)$$
$$\leq d(x, x_n) + d(x_n, y_n) + d(y_n, y)$$

so

$$d(x, y) - d(x_n, y_n) \leq d(x, x_n) + d(y_n, y).$$

Interchanging (x, y) and (x_n, y_n) gives

$$d(x_n, y_n) - d(x, y) \leq d(x, x_n) + d(y_n, y)$$

so

$$|d(x, y) - d(x_n, y_n)| \leq d(x, x_n) + d(y_n, y) \to 0.$$

∎

PROPOSITION 1.2

A convergent sequence has only one limit point.

PROOF   Assume that (x_n) has two limit points, x_1 and x_2.
Then $d(x_1, x_2) \leq d(x_1, x_n) + d(x_n, x_2) \to 0$ so $d(x_1, x_2) = 0$; hence
$x_1 = x_2$. ∎

Following the properties of the real numbers we define the concept of a
Cauchy sequence in a metric space

DEFINITION 1.4   *A sequence* (x_n) *in the metric space* (M, d) *is denoted a* Cauchy sequence *if* $d(x_n, x_m) \to 0$ *for* $n, m \to \infty$.

REMARK 1.1   In perhaps more familiar notation, this is written

$$\forall \varepsilon > 0 \quad \exists N : n, m > N \Rightarrow d(x_n, x_m) < \varepsilon. \tag{1.7}$$

∎

In R with the metric $d(x, y) = |x - y|$ we know that any Cauchy sequence
is convergent. We shall see below that this is an exclusive property of a
metric space. On the other hand, all convergent sequences are Cauchy:
$d(x_n, x_m) \le d(x_n, x) + d(x, x_m) \to 0$ if $\lim_n x_n = x$.

Example 1.5
Consider in the metric space $(C([-1; 1]), d_1)$ the sequence (f_n)shown below:

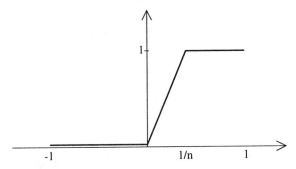

FIGURE 1.2
Graph of the nth element of (f_n)

It is obvious that

$$d_1(f_n, f_m) = \int_{-1}^{1} |f_n(t) - f_m(t)| \, dt$$

$$\le \frac{2}{\min\{n, m\}} \to 0$$

hence, (f_n) is a Cauchy sequence. But it is also obvious that (f_n) cannot
converge to any continuous function since the only possible candidate f

must satisfy

$$f(t) = \begin{cases} 1 \text{ for } & t \in \,]0,1] \\ 0 \text{ for } & t \in [-1,0[, \end{cases}$$

and it is not possible to extend f to a continuous function on $[-1;1]$.   ⬜

Since the property that all Cauchy sequences are convergent is exclusive, we have the following definition:

DEFINITION 1.5   *A metric space (M,d) is* complete *if all Cauchy sequences are convergent in (M,d).*

Example 1.6
We know that (R,d) with $d(x,y) = |x-y|$ is complete.
   We saw that $(C([a;b]),d_1)$ was not complete, but we know from elementary analysis that $(C([a;b]),d_\infty)$ is complete. The spaces $(C([a;b]),d_p)$ for $p > 1$ are not complete, either (an argument similar to the case $p = 1$ will show).   ⬜

There is a way to overcome the fact that some spaces fail to be complete - we make them so! This is not the whole truth, but the following will explain what one does. First we need the concept of a set being *dense* in another.

DEFINITION 1.6   *Let (M,d) be a metric space. A set $A \subset M$ is said to be* dense *in another set $B \subset M$ if, for all $b \in B$ and all $\varepsilon > 0$, there is an $a \in A$ with $d(a,b) < \varepsilon$.*

REMARK 1.2   Another way of saying this is that given $b \in B$, there is a sequence $(a_n) \subset A$ with $\lim_n a_n = b$.   ∎

Example 1.7
In (R,d) from example 1.6, Q is dense in R, and $]0;1[$ is dense in $[0;1]$.
⬜

Example 1.8
Consider the set P of polynomials on a bounded interval $[a;b]$. Obviously $P \subset C([a;b])$. Moreover, in the metric space $(C([a;b]),d_\infty)$, P is in fact dense in $C([a;b])$. This is the *Weierstrass Approximation Theorem*, and let's give a proof.

First notice that we can assume that the continuous function f we want to approximate is real since we can approximate real and imaginary parts one at a time. Moreover, we assume that $[a;b] = [0;1]$; this is just a scaling.

Define the sequence of polynomials $(p_n(f))$:

$$p_n(f)(t) = \sum_{k=0}^{n} f\left(\frac{k}{n}\right)\binom{n}{k} t^k (1-t)^{n-k}.$$

These are the so-called "Bernstein polynomials". Notice that f is uniformly continuous (since the interval $[0;1]$ is closed), that is,

$$\forall \varepsilon > 0 \quad \exists \delta > 0 \forall s, t \in [0;1] : |s - t| < \delta \Rightarrow |f(s) - f(t)| < \varepsilon.$$

Let $\varepsilon > 0$ be given and choose $\delta > 0$ according to this. For $t \in [0;1]$ we have:

$$|f(t) - p_n(t)| = \left| f(t)\left(t + (1-t)\right)^n - \sum_{k=0}^{n} f\left(\frac{k}{n}\right)\binom{n}{k} t^k (1-t)^{n-k} \right|$$

$$= \left| \sum_{k=0}^{n} \left(f(t) - f\left(\frac{k}{n}\right)\right)\binom{n}{k} t^k (1-t)^{n-k} \right|$$

$$\leq \sum_{|t-\frac{k}{n}|<\delta} \left| \left(f(t) - f\left(\frac{k}{n}\right)\right)\binom{n}{k} t^k (1-t)^{n-k} \right|$$

$$+ \sum_{|t-\frac{k}{n}|\geq\delta} \left| \left(f(t) - f\left(\frac{k}{n}\right)\right)\binom{n}{k} t^k (1-t)^{n-k} \right|$$

$$\leq \varepsilon \sum_{k=0}^{n} \binom{n}{k} t^k (1-t)^{n-k} + \sum_{\frac{\left(t-\frac{k}{n}\right)^2}{\delta^2}\geq 1} 2M \binom{n}{k} t^k (1-t)^{n-k},$$

where $M = \sup_{t\in[0;1]} |f(t)|$. Hence,

$$|f(t) - p_n(f)(t)| \leq \varepsilon + \frac{2M}{\delta^2} \sum_{k=0}^{n} \left(t - \frac{k}{n}\right)^2 \binom{n}{k} t^k (1-t)^{n-k}$$

$$= \varepsilon + \frac{2M}{\delta^2} \cdot \frac{1}{n} t(1-t)$$

since the expression in the last summation is just the variance of the binomial distribution.

Then

$$d_\infty(f, p_n(f)) \leq \varepsilon + \frac{2M}{\delta^2} \cdot \frac{1}{n} \cdot \frac{1}{4}$$

$$\leq 2\varepsilon \quad \text{for} \quad n > \frac{M}{\delta^2 \varepsilon},$$

and the result follows.   □

Now, continuing on the original path, we need the concept of *isometric spaces*.

DEFINITION 1.7   *The metric spaces* (M, d) *and* (M', d') *are* isometric *if there is a bijective map* $T : M \mapsto M'$ *such that* $d'(Tx, Ty) = d(x, y)$ *for all* $x, y \in M$.

THEOREM 1.3 (Completeness)
Let (M, d) *be a metric space. There is a complete metric space* (M', d') *and a dense set* $\tilde{M} \subset M'$ *such that* (M, d) *and* $(\tilde{M}, d')$ *are isometric.*
  The space (M', d') *is denoted the* **completion** *of* (M, d).

We will omit the proof but mention that all completions of a metric space are isometric; this is why we talk of *the* completion. Following this idea, that is to identify isometric spaces, we usually consider the original space as embedded in the completion. The completion of (Q, d) is (R, d) (where d is the natural metric) and we usually think of Q itself as a dense subset of R, even though by the construction of R it is an isometric image $T(Q)$ that is dense in R. The plot is that we are not able to distinguish between Q and $T(Q)$. Such identifications are made all the time in mathematics, and this book follows this tradition.

PROPOSITION 1.4
  $(C([a; b]), d_\infty)$ *is the completion of the space* (P, d_∞).

PROOF
  P is dense in $C([a; b])$, which is complete with respect to the metric d_∞.
∎

We will now prove the famous *Banach fixed point theorem*, and for this purpose we must define a certain class of mappings between metric spaces.

DEFINITION 1.8   *A mapping f from a metric space (M, d) into itself is called a* contraction *if there is an* $\alpha \in]0; 1[$ *such that*

$$d(f(x), f(y)) \leq \alpha d(x, y)$$

for all $x, y \in M$.
  We say that $x \in M$ is a fixed point *for f if $f(x) = x$.*

THEOREM 1.5 (Banach fixed point theorem)
Let (M, d) be a complete metric space and let f be a contraction. Then f has exactly one fixed point.

PROOF   Let α be the contraction constant.
 Assume that x and y are fixed points. Then

$$d(x, y) = d(f(x), f(y)) \leq \alpha d(x, y)$$

so $d(x, y) = 0$ since $\alpha < 1$. Hence f has at most one fixed point.
 Now choose $x_0 \in M$ and define the sequence (x_n) by $x_n = f^n(x_0)$. This is a Cauchy sequence since

$$\begin{aligned} d(x_{m+1}, x_m) &\leq d\left(f(x_m), f(x_{m-1})\right) \\ &\leq \alpha d(x_m, x_{m-1}) \\ &\cdots \\ &\leq \alpha^m d(x_1, x_0). \end{aligned}$$

Then

$$\begin{aligned} d(x_{n+p}, x_n) &\leq d(x_{n+p}, x_{n+p-1}) + \cdots + d(x_{n+1}, x_n) \\ &\leq \left(\alpha^{n+p-1} + \cdots + \alpha^n\right) d(x_1, x_0) \\ &= \alpha^n \frac{1 - \alpha^p}{1 - \alpha} d(x_1, x_0). \end{aligned}$$

Since $0 < \alpha < 1$, we have

$$d(x_{n+p}, x_n) \leq \frac{\alpha^n}{1 - \alpha} d(x_1, x_0),$$

hence (x_n) is a Cauchy sequence and is convergent to some $x \in M$. We apply the triangle inequality to show that $f(x) = x$:

$$\begin{aligned} d\left(f(x), x\right) &\leq d\left(f(x), x_n\right) + d(x_n, x) \\ &= d\left(f(x), f(x_{n-1})\right) + d(x_n, x) \\ &\leq \alpha d(x, x_{n-1}) + d(x_n, x) \to 0 \end{aligned}$$

so $d\left(f(x), x\right) = 0$.   ∎

REMARK 1.3   For some important applications of the fixed point theorem to integral equations, see Exercise 23.   ∎

 One can show that it is enough that f^m is a contraction for some m for f to have a unique fixed point, see Exercise 21.

It is also interesting in some applications to notice that the fixed point
theorem applies if f is a contraction on some closed, f-invariant subset
$A \subset M$ (That A is f-invariant means that $f(A) \subset A$).

Example 1.9

For a particularly simple application of the fixed point theorem, consider
the integral equation

$$x(t) - \mu \int_a^b k(t,s)x(s)ds = u(t),$$

where $u \in C([a;b])$, $a < b$, $k \in C([a;b]^2)$ are known functions, $\mu \in C$ is a
parameter, and x is the unknown function to be determined. To apply the
fixed point theorem, define $f : C([a;b]) \to C([a;b])$ by

$$f(x) = u + \mu \int_a^b k(\cdot,s)x(s)ds,$$

and notice that solving the equation is equivalent to finding the fixed point
for f in the complete metric space $(C([a;b]), d_\infty)$.

We will investigate when f is a contraction in the given metric.

Let $x, y \in C([a;b])$. For $t \in [a;b]$ we have

$$|f(x)(t) - f(y)(t)| = |\mu|| \int_a^b k(t,s)(x(s) - y(s))ds|$$

$$\leq |\mu| \int_a^b |k(t,s)|ds \cdot d_\infty(x,y)$$

$$\leq |\mu|M(b-a) \cdot d_\infty(x,y),$$

where $M = \sup_{t,s \in [a;b]} |k(t,s)|$.

This shows that

$$d_\infty(f(x), f(y)) \leq |\mu|M(b-a)d_\infty(x,y),$$

so f is a contraction when

$$|\mu| < \frac{1}{M(b-a)}.$$

When this is the case, we can solve the integral equation by the iteration

$$x_{n+1} = f(x_n) = u + \mu \int_a^b k(\cdot,s)x_n(s)ds,$$

starting with *any* $x_0 \in C([a;b])$, and converging to the unique, continuous
solution.   ▯

Chapter 2

Banach Spaces

The fundamental objects to be studied in functional analysis are functions, and since we can, in general, make linear combinations of functions, we are led into the framework of vector spaces. Furthermore, we would like to consider continuous maps between vector spaces, so these must have a topology. The vector spaces considered in this book are complex unless otherwise stated.

2.1 Normed Vector Spaces

DEFINITION 2.1 *Let V be a vector space. A* norm *on V is a function $\|\cdot\| \to [0; \infty[$ satisfying for all $x, y \in V$ and $\alpha \in C$:*

$$\|x + y\| \leq \|x\| + \|y\| \tag{2.1}$$
$$\|\alpha x\| = |\alpha|\|x\| \tag{2.2}$$
$$\|x\| = 0 \Rightarrow x = 0. \tag{2.3}$$

If the last condition above is omitted from the definition, $\|\cdot\|$ is called a *semi norm*.

We notice that any normed vector space is a metric space when we use the *norm induced metric* $d(x, y) = \|x - y\|$. Notice also that any norm is continuous, since $|\|x\| - \|y\|| \leq \|x - y\|$.

DEFINITION 2.2 *A normed vector space that is complete in the metric induced by the norm is called a* Banach space.

We have already noticed that $C([a; b)]$ with the *sup*-norm

$$\|f - g\|_\infty = d_\infty(f, g) \tag{2.4}$$

is a Banach space.

2.2 L^p-spaces

We will now construct some very important Banach spaces of functions, namely the L^p-spaces. These will be constructed from a suitable subspace of the continuous functions by imposing certain norms, and then completing the normed (hence metric) spaces.

First we need some definitions.

DEFINITION 2.3   *A set S in a normed (or just topological) space V is called* compact *if every sequence in S has a subsequence that converges to an element of S.*

Notice here that perhaps the simplest compact sets in R are the closed and bounded intervals.

DEFINITION 2.4   *The* support *of a function $f : R \to C$ is the closure of the set $\{x \in R \mid f(x) \neq 0\}$, usually denoted supp$(f)$. The space $C_0(R)$ is the space of continuous functions with bounded, hence compact, support.*

REMARK 2.1   A continuous function defined on a closed, bounded interval can be considered as a function with compact support by a trivial extension.   ∎

We now equip $C_0(R)$ with the *p-norms*, that is, the norms that induce the d_p-metrics from the previous chapter.

PROPOSITION 2.1
The map $\| \cdot \|_p$, , from $C_0(R)$ to R, $p \geq 1$, given by

$$\|f\|_p = (\int_R |f(x)|^p dx)^{\frac{1}{p}},$$

is a norm on $C_0(R)$.

PROOF   The only condition that is trivial is (2.2).

Let us first verify (2.3).

Assume that $\|f\|_p = 0$ and $f \neq 0$. Then there is an x_0 such that $|f(x_0)| > 0$. By the continuity of f, we can find a $\delta > 0$ such that, say,

$$|f(x) - f(x_0)| < \frac{1}{3}|f(x_0)| \quad \text{for} \quad |x - x_0| < \delta.$$

Then $|f(x)| > \frac{1}{3}|f(x_0)|$ for $|x - x_0| < \delta$, so we have that

$$\|f\|_p = (\int_R |f(x)|^p dx)^{\frac{1}{p}} \geq (2\delta)^{\frac{1}{p}} \cdot \frac{1}{3}|f(x_0)| > 0,$$

a contradiction.   ∎

The proof of (2.1), the triangle inequality, is rather long and consists of several steps, each of independent interest. We will therefore state the steps separately but keep the goal in mind. The case $p = 1$ is, of course, trivial.

LEMMA 2.2

Young's inequality

Let $p > 1, q > 1$ and $\frac{1}{p} + \frac{1}{q} = 1$, and let $a > 0$ and $b > 0$. Then

$$ab \leq \frac{1}{p}a^p + \frac{1}{q}b^q.$$

PROOF

We consider the graph of $y = x^{p-1}$:

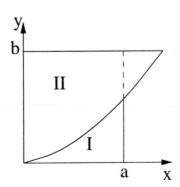

FIGURE 2.1

Graph of $y = x^{p-1}$.

Notice that $x = y^{q-1}$, hence the area of the rectangle is ab, the area of I is $\frac{1}{p}a^p$, and the area of II is $\frac{1}{q}b^q$. The inequality follows. ∎

LEMMA 2.3
Hölder's inequality, continuous case
Let $p > 1, q > 1$ and $\frac{1}{p} + \frac{1}{q} = 1$, and assume that $f, g \in C_0(R)$.
Then

$$\|fg\|_1 \le \|f\|_p\|g\|_q.$$

PROOF   The statement is obviously true for $f = 0$ or $g = 0$, so assume that $\|f\|_p > 0$ and $\|g\|_q > 0$. Then

$$\frac{|f(x)g(x)|}{\|f\|_p\|g\|_q} \le \frac{1}{p}\frac{|f(x)|^p}{\|f\|_p^p} + \frac{1}{q}\frac{|g(x)|^q}{\|g\|_q^q},$$

hence

$$\int_R \frac{|f(x)g(x)|}{\|f\|_p\|g\|_q}dx \le \frac{1}{p} + \frac{1}{q} = 1,$$

and the inequality follows.   ∎

REMARK 2.2   The special case of Hölder's inequality where $p = q = 2$ is called *Cauchy-Schwarz' inequality*. It will be stated precisely later. ∎

Now we are ready to prove (2.1), the triangle inequality

LEMMA 2.4
Minkowski's inequality, continuous case
Let $p \ge 1$ and assume that $f, g \in C_0(R)$.
Then

$$\|f + g\|_p \le \|f\|_p + \|g\|_p.$$

PROOF   The case $p = 1$ is trivial. For $p > 1$, write $\frac{1}{q} = 1 - \frac{1}{p}$ and see that

$$\int_R |f(x) + g(x)|^p dx \le \int_R |f(x)||f(x) + g(x)|^{p-1}dx$$

$$+ \int_R |g(x)||f(x) + g(x)|^{p-1}dx$$

$$\le \|f\|_p\|(f + g)^{(p-1)}\|_q + \|g\|_p\|(f + g)^{(p-1)}\|_q,$$

according to Hölder's inequality. Since $q(p-1) = p$, the last expression equals

$$(\|f\|_p + \|g\|_p)(\int_R |f(x) + g(x)|^p dx)^{1-\frac{1}{p}}.$$

Then, dividing both sides of the inequality with the last factor, one gets

$$(\int_R |f(x) + g(x)|^p dx)^{\frac{1}{p}} \leq \|f\|_p + \|g\|_p,$$

which is the inequality. (If the last factor is zero, the inequality is, of course, trivial.) ∎

Now we have shown that $C_0(R)$ with the p-norms are normed spaces.

REMARK 2.3   Any complex sequence (x_n) can be considered as a function $f : N \to C$ where we define $f(n) = x_n$. The natural way to translate the above into results about sequences is to let $C_0(N)$ be the space of sequences with only a finite number of elements different from zero. The inequalities above are then valid, with the integrals substituted with summations.

Notice also that a similar identification allows us, for a vector $x \in C^k$, to speak of its p-norm; in particular, the 2-norm of a vector is denoted the *Euclidean norm*. ∎

As we saw in example (1.5), the spaces $C_0(R)$ with the p-norms are not complete. We will now complete them as metric spaces. This is not a trivial construction, and we will comment on it after the following definition.

DEFINITION 2.5   *The vector space $L^p(R)$; $p \geq 1$ is the completion of $C_0(R)$ in the metric induced by the p-norm.*

The L^p spaces are now Banach spaces *by construction*. It is *not* evident what these spaces really consist of. Take for example a continuous function f, satisfying $\int_R |f(x)|^p < \infty$. Then $f_n = f \cdot \chi_{[-n;n]}$ is in $C_0(R)$, for all n, where $\chi_{[-n;n]}$ is the function with the graph shown in Figure 2.2. We see that

$$\int_R |f(x) - f_n(x)|^p dx \leq \int_{-\infty}^{-n} |f(x)|^p dx + \int_n^{\infty} |f(x)|^p dx \to 0 \quad \text{for} \quad n \to \infty,$$

since $\int_R |f(x)|^p dx$ is convergent.

Hence $f_n \to f$ in p-norm and f is in $L^p(R)$. But if we consider the discontinuous function f_1 constructed from f by changing the value in one single point (see Figure 2.3),

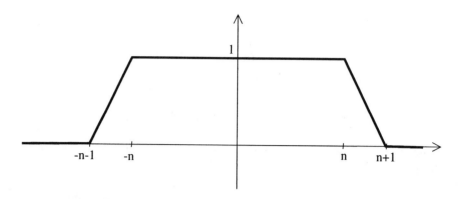

FIGURE 2.2

Figure of $\chi_{[-n;n]}$

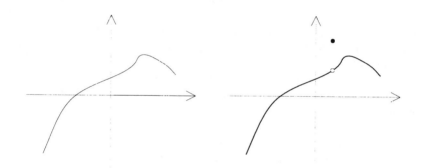

FIGURE 2.3

Figures of f and f_1

we see that

$$\int_R |f(x) - f_1(x)|^p dx = 0,$$

so the convergent sequence (f_n) converges also to f_1, so this function must also belong to $L^p(R)$ - but we have now the obvious problem that in this space $f - f_1$ must be the zero function !! This is the price we have to pay if we want our p-normed spaces to be complete.

Functions cannot in general be considered, pointwise (of course, for the good old continuous ones there are no problems, according to Proposition 2.1). This can be put into a consistent framework using the *Lebesgue theory of integration,* which is not within the scope of this book. Instead of functions, one speaks of *equivalence classes* of functions, and two functions are said to be *equivalent* if their difference is the zero function in $L^p(R)$. In this case we will write

$$f = g \quad a.e, \tag{2.5}$$

where *a.e* means *almost everywhere*. But the integral defining the norm is now the so-called *Lebesgue integral*. That allows us to integrate a much larger class of functions than the Riemann integrable ones. For example, 1_Q, the *characteristic function* for Q that takes the value 1 if $x \in Q$ and is 0 elsewhere, is Lebesgue integrable with integral 0, whereas it is not Riemann integrable over any interval. But for the continuous functions there are no problems; if f is continuous, then f belongs to $L^p(R)$ if and only if $\int_R |f(x)|^p dx < \infty$. We must confine ourselves to the fact that there are many more *p-integrable functions*, but we can *always* rely on the denseness of $C_0(R)$ in $L^p(R)$, so we can approximate with nice, continuous, and compactly supported functions. We will refer to [26] for a comprehensive introduction to integration theory.

All this can of course be done when R is replaced by R^k and the spaces $L^p(R^k)$ are defined in the obvious way. Moreover, if R is replaced by any interval I, the spaces $L^p(I)$ are defined similarly.

This gives us also the precise formulations of Hölder's, Minkowski's, and Cauchy-Schwarz' inequalities:

THEOREM 2.5 (Hölders inequality)
Let $p > 1, q > 1$ and $\frac{1}{p} + \frac{1}{q} = 1$, and assume that $f \in L^p(R)$ and $g \in L^q(R)$.
Then

$$\|fg\|_1 \leq \|f\|_p \|g\|_q.$$

THEOREM 2.6 (Cauchy-Schwarz' inequality)
Assume that $f, g \in L^2(R)$.
Then

$$\|fg\|_1 \leq \|f\|_2 \|g\|_2.$$

THEOREM 2.7 (Minkowski's inequality)
Let $p \geq 1$ and assume that $f, g \in L^p(R)$.
Then

$$\|f + g\|_p \leq \|f\|_p + \|g\|_p.$$

It will be vital for us to be able to approximate in L^p, so we will find some particularly simple spaces that are dense.

DEFINITION 2.6 *The space of functions that are infinitely many times differentiable (smooth) and compactly supported is called $C_0^\infty(R)$. The* space $C_0^\infty(I)$, where I is an interval, is defined analogously. Moreover, $P(R)$ and $P(I)$ denote the spaces of polynomials on R or I, respectively.

REMARK 2.4   One could wonder if such C_0^∞-functions exist. Consider the function

$$\varphi_{a,b}(x) = \begin{cases} e^{\frac{-1}{(x-a)(b-x)}} & \text{for} \quad x \in]a; b[; \\ 0 & \text{otherwise.} \end{cases}$$

This is a $C_0^\infty(R)$-function .   ∎

THEOREM 2.8
(i) $C_0^\infty(R)$ is dense in $C_0(R)$, both in the sup-norm and the p-norm.
(ii) $C_0^\infty(R)$ is dense in $L^p(R)$.
(iii) $C_0^\infty(I)$ is dense in $L^p(I)$.
(iv) If I is finite, $P(I)$ is dense in $L^p(I)$.
(v) $C_0^\infty(I)$ is dense in $C(I)$ with respect to the p-norm.

PROOF   (i): We define first a so-called *mollifier* m_α as follows: For each $\alpha > 0$, m_α is a smooth, nonnegative function with $m_\alpha(x) = 0$ for $|x| > \alpha$ and $\int_R m_\alpha(x)dx = 1$. (Just think of $\varphi_{a,b}$ above, with $a = -\alpha$, $b = \alpha$, and adjusted with a constant in order to make the integral $= 1$.)
Now, for $f \in C_0(R)$ we define the function f_α by

$$f_\alpha(x) = \int_R f(x - y)m_\alpha(y)dy,$$

and notice that by a change of variable we have

$$f_\alpha(x) = \int_R f(z)m_\alpha(x - z)dz,$$

so f_α is in $C_0^\infty(R)$ since m_α is.
Since f is compactly supported it is uniformly continuous, so given $\epsilon > 0$ we can find $\alpha > 0$, such that

$$|f(x - y) - f(x)| < \epsilon \quad \text{for} \quad |y| < \alpha.$$

Then

$$|f_\alpha(x) - f(x)| = |\int_R (f(x - y) - f(x))m_\alpha(y)dy| \tag{2.6}$$

$$\leq \int_R |f(x - y) - f(x)|m_\alpha(y)dy \tag{2.7}$$

$$< \epsilon, \tag{2.8}$$

due to the properties of the mollifier m_α.

Hence $\|f_\alpha - f\|_\infty < \epsilon$ and $C_0^\infty(R)$ is dense in $C_0(R)$ with the sup-norm. But observe that since f and f_α are compactly supported, we can find $a, b \in R$ such that

$$\|f_\alpha - f\|_p^p = \int_R |f_\alpha(x) - f(x)|^p dx \le \int_a^b \epsilon^p dx = \epsilon^p(b-a),$$

so

$$\|f_\alpha - f\|_p \le \epsilon(b-a)^{\frac{1}{p}},$$

showing that $C_0^\infty(R)$ is also dense in $C_0(R)$ with the p-norm.

(ii): It is obvious that if A is dense in B, and B dense in C, then A is dense in C. From (i), $C_0^\infty(R)$ is dense in $C_0(R)$, which (by definition) is dense in $L^p(R)$, and the result follows.

The proof of *(v)* is omitted, but it is a variant of the proof of *(i)*. Now *(iii)* follows directly from *(v)*. Moreover, *(iv)* follows directly from the Weierstrass' approximation theorem. ∎

Following the point of view from Remark 2.2.3, it is natural to define the analogue sequence spaces to the $L^p(R)$ spaces, denoted l^p.

DEFINITION 2.7 *The space of sequences (real or complex) l^p, $p \ge 1$, consists of sequences (x_n) satisfying*

$$\sum_{n=1}^\infty |x_n|^p < \infty.$$

The norm in l^p is

$$\|(x_n)\|_p = (\sum_{n=1}^\infty |x_n|^p)^{\frac{1}{p}}.$$

The proper translation of Theorem 2.8 is that the space of sequences with only a finite number of elements different from zero is dense in l^p for all p.

2.3 Infinite Dimensional Spaces

The Minkowski inequality shows, among other things, that the $L^p(R)$ spaces *are* in fact vector spaces, since they are stable under the vector space operations. This justifies the terminology we have been using from the very beginning. But since they differ quite a lot from the well-known finite dimensional ones, we will take a glance at how the basic definitions look.

Let $x_1, x_2, ..., x_k$ be elements in the vector space V. Recall that this set of vectors is called *linearly independent* if the equation

$$\alpha_1 x_1 + \alpha_2 x_2 + ... + \alpha_k x_k = 0$$

is only satisfied for $\alpha_1 = \alpha_2 = ... = \alpha_k = 0$. If this is not the case, the set is called *linearly dependent*. Notice also that any subset of a linearly independent set of vectors is again linearly independent. Moreover, a vector $x \in V$ is called a *linear combination* of the vectors $x_1, x_2, ..., x_k$ if there are constants $\alpha_1, \alpha_2, ..., \alpha_k$, such that $\alpha_1 x_1 + \alpha_2 x_2 + ... + \alpha_k x_k = x$.

Let U be a subset of V. By $span(U)$ we denote the vector space consisting of all finite linear combinations of vectors in U. This is of course a subspace of V. If the vector space V contains a finite set of k linearly independent vectors $x_1, x_2, ..., x_k$, and moreover any set of $k + 1$ vectors is linearly dependent, we will say that the *dimension* of V is k, and write $dim(V) = k$. It follows that any vector $x \in V$ is then a linear combination $x = \alpha_1 x_1 + \alpha_2 x_2 + ... + \alpha_k x_k$ of at most k linearly independent vectors. If there is no such $k \in N$, we will say that V is *infinite dimensional*.

PROPOSITION 2.9

Any vector space containing $C_0^\infty(R)$ is infinite dimensional.

PROOF   Consider the set of functions $\{e_n\}_{n \in Z} \subset C_0^\infty(R)$, where e_n is defined in the following way:

$$e_n(x) = \varphi_{n,n+1}(x) = \begin{cases} e^{\frac{-1}{(x-n)(n+1-x)}} & \text{for} \quad x \in]n; n+1[; \\ 0 & \text{otherwise.} \end{cases}$$

This is clearly an infinite set with the property that any finite subset is linearly independent, hence $\{e_n\}_{n \in Z}$ is infinite dimensional. Then $C_0^\infty(R)$ is also infinite dimensional, and so is any vector space containing it.   ∎

Now the vector space operations allow us to define what is meant by a *series*.

DEFINITION 2.8   *Let $x_1, x_2, ...$ belong to the normed vector space V. We say that the series $x_1 + x_2 + ...$ converges, if there is an $x \in V$, such that the sequence (s_n) of partial sums $s_n = \sum_{i=1}^{n} x_i$ converges to x. When this is the case, we write $x = \sum_{i=1}^{\infty} x_i$.*

This opens up for the precise definition of a *basis* for a normed vector space V. If the dimension of V is finite, and $dim(V) = k$, then any vector $x \in V$ is a linear combination of at most k linearly independent vectors from V, and we know from linear algebra that any set U of k linearly independent vectors from V is a basis for V. In this case we have that $span(U) = V$. Recall that the dimension of V equals the number of vectors in the basis. If the dimension of V is infinite and we furthermore assume that V is a Banach space, we will say that a set U of linearly independent vectors from V is *total* in V if any $x \in V$ is the limit of a sequence (s_n) of vectors $s_n \in span(U)$. Then $V = \overline{span}(U)$, and we see that any vector $x \in V$ can be expressed as a series $x = \lim_n s_n = \lim_n \sum_{i=1}^{n} \alpha_i x_i = \sum_{i=1}^{\infty} \alpha_i x_i$ of vectors $x_i \in U$, $\alpha_i \in C$. We will call the total set U a (Schauder) *basis* , if all $x \in H$ have such a *unique* series expression. A normed vector space is called *separable* if it contains a countable, dense subset, and we are only concerned about separable spaces in this book.

Chapter 3

Bounded Operators

A mapping from a normed space V into another normed space W is called an *operator*. If the mapping is linear, that is, if

$$T(\alpha x + \beta y) = \alpha Tx + \beta Ty,$$

for all α, β in C and x, y in V, we call T a *linear operator*. We will sometimes consider operators defined only on a subset of V, called the *domain* of T and usually denoted $D(T)$. If so, in order for T to be linear, we will also demand that $D(T)$ is a subspace of V. In linear algebra we have studied the linear operators from C^n into C^k, the matrices. The aim is now to extend the setting to study continuity properties of operators when the spaces are only assumed to be normed. Since normed spaces are topological spaces with the topology defined by the norms, the definition of continuity can now be written in a perhaps more intuitive manner.

3.1 Basic Properties

DEFINITION 3.1   *An operator T from a normed space V into another normed space W is* continuous *at a point $x \in D(T)$ if for any $\epsilon > 0$ there is a $\delta > 0$, such that $\|Tx - Ty\| < \epsilon$ for all $y \in D(T)$ with $\|x - y\| < \delta$. T is* continuous *if it is continuous at all points of $D(T)$.*

T is uniformly continuous *on $D(T)$ if for any $\epsilon > 0$ there is a $\delta > 0$, depending only on ϵ, such that $\|Tx - Ty\| < \epsilon$ whenever $\|x - y\| < \delta$.*

T is bounded *if there is a positive constant M such that $\|Tx\| \leq M\|x\|$ for all $x \in D(T)$.*

This is the obvious extension of the concept of continuity from real analysis, saying that objects close together are mapped into objects close to-

gether. Continuous real functions in the usual sense are continuous opera-
tors in the sense above. Notice how the norm symbol $\|\cdot\|$ is used, denoting
both the V-norm and the W-norm. In applications it is important to keep
in mind in which space one is working, but it is a significant power of
abstract mathematics to be able to write simple formulas and definitions.

Now we will show that continuity for linear operators is equivalent to
boundedness.

PROPOSITION 3.1
*Let T be a linear operator from a normed space V into a normed space W.
The following statements are equivalent:*
(i): T is continuous in 0.
(ii): T is continuous.
(iii): T is bounded.

PROOF   Assume that T is continuous in 0. Let $\epsilon > 0$ be given, then we
can find $\delta > 0$, such that $\|z\| < \delta$ implies that $\|Tz\| < \epsilon$. Hence $\|x - y\| < \delta$
implies that $\|Tx - Ty\| = \|T(x - y)\| < \epsilon$, so T is (uniformly) continuous.
     It is trivial that (ii) implies (i), so assume now that T is bounded.
     Then $\|Tx - Ty\| = \|T(x - y)\| \leq M\|x - y\|$, so T is continuous.
     On the other hand, assume that T is unbounded. Then we can find a
sequence (x_n) in V, such that $\|Tx_n\| \geq n\|x_n\|$ for all $n \in N$. Now define the
sequence (y_n) by $y_n = \frac{x_n}{n\|x_n\|}$ and notice that $\|Ty_n\| \geq 1$, in contradiction
to the fact that obviously $y_n \to 0$, hence T is not continuous in 0.   ∎

This shows that a linear operator is either everywhere continuous or
nowhere continuous.
     For a linear operator T we define the *kernel*, denoted $ker(T)$, of T as the
set $\{x \in D(T) \mid Tx = 0\}$. It is easy to see that this is a subspace of V, and
it follows that $ker(T)$ is closed whenever T is bounded.
     A matrix is a continuous operator (since it is obviously bounded in the
usual norms on C^n), so linear algebra does not provide us with an example
of an unbounded operator. But here is one :

Example 3.1
Let $f \in C^1(I)$, where I is a bounded interval. Consider the canonical example of a linear operator, namely differentiation: $Df = f'$. D is obviously
linear from the space $C^1(I)$ into $C^0(I)$, both equipped with the sup-norms.
Let (f_n) be the sequence of functions

$$f_n(x) = \sin(nx),$$

so that $\|f_n\|_\infty = 1$ and $\|Df_n\|_\infty = n$. This is an example of a sequence in the unit ball of C^1 that is mapped into an unbounded sequence, so D is unbounded.

But let us now choose the D-norm on C^1, defined by

$$\|f\|_D = \|f\|_\infty + \|Df\|_\infty.$$

Now $\|Df\|_\infty \leq \|f\|_D$, and D is now a bounded operator between two normed spaces. This is, of course, just an example of how to choose the right topology in order to make things work. ▯

Example 3.2
Another example of a continuous map is the norm itself. Let V be any normed space, then the norm is a map from V into R that satisfies

$$\big|\|x\| - \|y\|\big| \leq \|x - y\|,$$

hence $\|\cdot\| : V \to R$ is continuous (but not linear, of course). ▯

Before we continue with the study of the linear operators in particular, we will show how to *extend* a continuous operator.

DEFINITION 3.2 *Let T be an operator with domain $D(T)$, and T_1 an operator with domain $D(T_1)$. If $D(T) \subset D(T_1)$ and $Tx = T_1 x$ for all $x \in D(T)$, we call T_1 an* extension *of T, and T is a* restriction *of T_1.*

There is only one way to extend uniformly continuous operators, as we will now see.

THEOREM 3.2
Let T be a uniformly continuous operator from a normed space V into a Banach space W, and assume that $D(T)$ is dense in V. Then T has a unique, continuous extension $T_1 : V \to W$.

PROOF Let $x' \in V$ and choose a sequence (x_n) from $D(T)$ that converges to x'. Recall that (x_n) is then a Cauchy sequence. We will show that (Tx_n) is a Cauchy sequence and define its limit to be $T_1 x'$.

Because T is uniformly continuous, for any $\epsilon > 0$ there is a $\delta > 0$ such that $\|Tx_n - Tx_m\| < \epsilon$ whenever $\|x_n - x_m\| < \delta$. Because (x_n) is a Cauchy sequence, there is an n_0, such that $\|x_n - x_m\| < \delta$ for $n, m > n_0$. Hence

$$\|Tx_n - Tx_m\| < \epsilon \quad \text{for} \quad n, m > n_0,$$

showing that (Tx_n) is a Cauchy sequence in the Banach space W and therefore convergent, $Tx_n \to u$, say. We define $T_1 x' = u$. That is, for each $x' \in V$ we choose a sequence (x_n) in $D(T)$ with $x_n \to x'$ and define $T_1 x' = \lim_n (Tx_n)$. Since there are many sequences converging to x', we have to show that this defines T_1 uniquely. We will first show that T_1 is uniformly continuous.

Take $x', y' \in V$, and corresponding sequences $(x_n), (y_n)$ in $D(T)$ so that $x_n \to x'$ and $y_n \to y'$. Then

$$\|T_1 x' - T_1 y'\| = \|\lim_n (Tx_n - Ty_n)\| = \lim_n \|Tx_n - Ty_n\|.$$

Since T is uniformly continuous, given any $\epsilon > 0$ there is a $\delta > 0$ such that

$$\|Tx_n - Ty_n\| < \frac{\epsilon}{2} \quad \text{when} \quad \|x_n - y_n\| < \delta.$$

There is an n_0 such that $\|x_n - x'\| < \frac{\delta}{3}$ and $\|y_n - y'\| < \frac{\delta}{3}$ for $n > n_0$. Now, assume that $\|x' - y'\| < \frac{\delta}{3}$, then

$$\|x_n - y_n\| \leq \|x_n - x'\| + \|x' - y'\| + \|y_n - y'\| < \delta,$$

hence

$$\|Tx_n - Ty_n\| < \frac{\epsilon}{2} \quad \text{for} \quad n > n_0.$$

This shows that $\|T_1 x' - T_1 y'\| < \epsilon$ for $\|x' - y'\| < \frac{\delta}{3}$, and T_1 is uniformly continuous.

To see that T_1 is unique, assume that T_2 is also a continuous extension of T. For any $x' \in V$, take a sequence (x_n) in $D(T)$ converging to x'. Then

$$T_1 x' = \lim_n (T_1 x_n) = \lim_n (Tx_n) = \lim_n (T_2 x_n) = T_2 x',$$

so $T_1 = T_2$. ∎

3.2   Bounded Linear Operators

We will now return to the study of linear operators. Let V and W be two normed spaces, and let S and T be linear, bounded, and hence continuous operators from V into W. It is obvious that for $\alpha, \beta \in C$, the operator $\alpha S + \beta T$ is also bounded (since it is continuous), so the bounded operators from V into W form a vector space. This space we will denote $B(V, W)$. In the special case where $V = W$, we will write $B(V)$ instead of $B(V, V)$. We will now equip $B(V, W)$ with a norm, the so-called *operator norm*.

DEFINITION 3.3   *Let V and W be normed vector spaces. For $T \in B(V, W)$, we define the* norm *of T by*

$$\|T\| = \sup\{\|Tx\| \mid \|x\| \leq 1\}.$$

Notice how the norm symbol $\| \cdot \|$ is used for the norm in three different spaces. It is easy to see that $B(V, W)$ with the operator norm in fact is a normed space.

We observe that $\|Tx\| \leq \|T\|\|x\|$ for all $x \in V$, and that any $M > 0$ satisfying $\|Tx\| \leq M\|x\|$ for all $x \in V$ must also be $\geq \|T\|$.

Now, since the bounded operators form a vector space, we can consider sequences and series of bounded operators, and an important result is the following theorem.

THEOREM 3.3
Let V be a normed vector space and W a Banach space. Then $B(V, W)$ is a Banach space.

PROOF   Assume that (T_n) is a Cauchy sequence in $B(V, W)$. For $\epsilon > 0$ there is an n_0 such that $\|T_n - T_m\| < \epsilon$ for $n, m > n_0$. Then, for any $x \in V$ we have

$$\|T_n x - T_m x\| = \|(T_n - T_m)x\| \leq \|T_n - T_m\|\|x\|,$$

showing that $(T_n x)$ is a Cauchy sequence in W. So for any $x \in V$ we can define an operator T by $Tx = \lim_n T_n x$. This operator is obviously linear, since for all $\alpha, \beta \in C$, $x, y \in V$:

$$\begin{aligned}
T(\alpha x + \beta y) &= \lim_n (T_n(\alpha x + \beta y)) \\
&= \lim_n (\alpha T_n x + \beta T_n y) \\
&= \alpha \lim_n T_n x + \beta \lim_n T_n y \\
&= \alpha Tx + \beta Ty.
\end{aligned}$$

Now we must show that T is bounded, and that (T_n) converges to it.

Let $n > n_0$. From the inequality

$$\big|\|Tx\| - \|T_n x\|\big| \leq \|Tx - T_n x\| = \|\lim_m T_m x - T_n x\| = \lim_m \|T_m x - T_n x\|,$$

we see that

$$\|Tx\| \leq \|T_n x\| + \lim_m \|T_m x - T_n x\| \leq (\|T_n\| + \lim_m \|T_m - T_n\|)\|x\|.$$

Since (T_n) is a Cauchy sequence, it is bounded, and since $n > n_0$, we have that $\lim_m \|T_n - T_m\| < \epsilon$. This shows that $\|Tx\| \leq (K + \epsilon)\|x\|$, for some constant $K > 0$, hence T is bounded.

Since

$$\|Tx - T_n x\| = \|\lim_m T_m x - T_n x\| = \lim_m \|T_m x - T_n x\| \leq \lim_m \|T_m - T_n\|\|x\| \leq \epsilon \|x\|,$$

it is clear that $\|T - T_n\| \leq \epsilon$ for $n > n_0$, hence $T_n \to T$ in $B(V, W)$. ∎

The special case when the Banach space W is C calls for some comments. Here $B(V, W)$ is then the Banach space of bounded, linear operators from V into C. This is also called the space of bounded, linear *functionals* on V, or the *dual space* of V, frequently denoted $V^\star$.

Example 3.3

Assume that $p, q > 1$ and $\frac{1}{p} + \frac{1}{q} = 1$, and let $f \in L^p(R)$. We can define a linear, bounded functional $T_f \in L^q(R)^\star$ by

$$T_f g = \int_R f(x)g(x)dx,$$

for $g \in L^q(R)$. From Hölders inequality we see that

$$|T_f g| \leq \int_R |f(x)g(x)|dx \leq \|f\|_p \|g\|_q,$$

hence $\|T_f\| \leq \|f\|_p$.

This shows that any $f \in L^p(R)$ in a natural way defines a bounded, linear functional T_f on $L^q(R)$. (It can be shown that, in fact, $L^p(R) = L^q(R)^\star$.)
◻

If V, W, Z are normed spaces, and $S \in B(V, W), T \in B(W, Z)$, we can define the linear operator TS by $TSx = T(Sx)$. Then $TS \in B(V, Z)$, since

$$\|TSx\| \leq \|T\|\|Sx\| \leq \|T\|\|S\|\|x\|,$$

showing that $\|TS\| \leq \|T\|\|S\|$. If $T \in B(V)$, we can define the powers $T^n, n \in N$ of T; these are also bounded since successive applications of the inequality gives

$$\|T^n x\| \leq \|T^n\|\|x\| \leq \|T\|^n\|x\|,$$

which implies that $\|T^n\| \leq \|T\|^n$.

This fact gives us the very interesting next proposition.

PROPOSITION 3.4

Assume that V is a Banach space and let $\varphi . C \to C$ be a complex function given by a convergent power series

$$\varphi(z) = \sum_{n=0}^{\infty} a_n z^n \quad \text{for} \quad |z| < \rho,$$

where ρ is the radius of convergence for the series and $a_n \in C$. If $T \in B(V)$ and $\|T\| < \rho$, the series $\sum_{n=0}^{\infty} a_n T^n$ is convergent in $B(V)$. We define in this way the operator-valued function φ by

$$\varphi(T) = \sum_{n=0}^{\infty} a_n T^n.$$

Moreover, we have that $\|\varphi(T)\| \leq |\varphi(\|T\|)|$.

PROOF We use the convention $T^0 = I$, where I is the identity operator.

Define the sequence (S_n) in $B(V)$ by $S_n = \sum_{k=0}^{n} a_k T^k$. We only have to show that (S_n) is a Cauchy sequence.

Let $\epsilon > 0$. Since the radius of convergence is ρ, the real series $\sum_{n=0}^{\infty} |a_n| \|T\|^n$ is convergent for $\|T\| < \rho$, so there is a number $n_0 \in N$ such that

$$\sum_{n=k}^{\infty} |a_n| \|T\|^n < \epsilon \quad \text{for} \quad k > n_0.$$

For $n, p \in N$, $n > n_0$ we have

$$\|S_{n+p} - S_n\| = \left\| \sum_{k=n+1}^{n+p} a_k T^k \right\|$$

$$\leq \sum_{k=n+1}^{n+p} |a_k| \|T\|^k$$

$$\leq \sum_{k=n+1}^{\infty} |a_k| \|T\|^k$$

$$< \epsilon,$$

showing that (S_n) is a Cauchy sequence. This also shows the last statement in the proposition. ∎

This proposition has a number of nontrivial applications, many frequently used in numerical analysis. The following corollary defines the so-called *Neumann series* of an operator, and is very commonly used.

COROLLARY 3.5

Assume that $T \in B(V)$, where V is a Banach space. If $\|T\| < 1$, the operator $I - T$ has an inverse $(I - T)^{-1} \in B(V)$. Moreover, $(I - T)^{-1}$ is given by the Neumann series:

$$(I - T)^{-1} = \sum_{n=0}^{\infty} T^n. \tag{3.1}$$

PROOF

From proposition 3.4 it is obvious that the series is convergent in $B(V)$, so denote the limit S. Hence $S = \lim_n S_n = \lim_n \sum_{k=0}^{n} T^k$. But notice that

$$(I - T)S_n = (I - T)(I + T + T^2 + ... + T^n)$$
$$= I - T^{n+1}$$
$$= (I + T + T^2 + ... + T^n)(I - T)$$
$$= S_n(I - T),$$

where $I - T^{n+1} \to I$ for $n \to \infty$ because

$$\|I - (I - T^{n+1})\| \leq \|T^{n+1}\| \leq \|T\|^{n+1} \to 0 \quad \text{for} \quad n \to \infty.$$

Hence $(I - T)S = S(I - T)$ and the result follows.   ∎

Example 3.4

Another operator frequently met in applications is the exponential e^T, defined for all $T \in B(V)$, since the series for e^z converges for all $z \in C$. Notice that for any $s \in R$, the operator e^{Ts} is also in $B(V)$, so if $x \in V$, we can define a mapping $u : R \to V$ by

$$u(s) = e^{Ts}x.$$

In the special case where $V = R^k$, the bounded, linear operators are just the matrices, and the u above is the usual way to express the solution of the ordinary differential matrix equation

$$\frac{du}{ds} = Tu$$
$$u(0) = x,$$

and, without much effort, this can be generalized to infinite dimensional Banach spaces also.   ∎

Example 3.5

Let us consider $V = R^2$ and the linear operator represented by the matrix

$$T_\theta = \begin{pmatrix} 0 & \theta \\ -\theta & 0 \end{pmatrix}$$

Simple calculations show that

$$T_\theta^{2n} = (-1)^n \theta^{2n} \begin{pmatrix} 1 & 0 \\ 0 & 1 \end{pmatrix}, \quad T_\theta^{2n+1} = (-1)^n \theta^{2n+1} \begin{pmatrix} 0 & 1 \\ -1 & 0 \end{pmatrix},$$

hence

$$e^{T_\theta} = \sum_{n=0}^{\infty} \frac{1}{n!} T_\theta^n$$

$$= \sum_{n=0}^{\infty} \frac{1}{(2n)!} T_\theta^{2n} + \sum_{n=0}^{\infty} \frac{1}{(2n+1)!} T_\theta^{2n+1}$$

$$= \left(\sum_{n=0}^{\infty} \frac{(-1)^n}{(2n)!} \theta^{2n}\right) \begin{pmatrix} 1 & 0 \\ 0 & 1 \end{pmatrix} + \left(\sum_{n=0}^{\infty} \frac{(-1)^n}{(2n+1)!} \theta^{2n+1}\right) \begin{pmatrix} 0 & 1 \\ -1 & 0 \end{pmatrix}$$

$$= \cos(\theta) \begin{pmatrix} 1 & 0 \\ 0 & 1 \end{pmatrix} + \sin(\theta) \begin{pmatrix} 0 & 1 \\ -1 & 0 \end{pmatrix}$$

$$= \begin{pmatrix} \cos(\theta) & \sin(\theta) \\ -\sin(\theta) & \cos(\theta) \end{pmatrix}.$$

This shows that e^{T_θ} is a rotation in R^2 with the angle of rotation $-\theta$. This is of course just linear algebra, but here we have no problems with convergence of the series $\sum_{n=0}^{\infty} \frac{1}{n!} T_\theta^n$.

□

We will conclude this section with two famous theorems that will be stated without proofs.

THEOREM 3.6 (Hahn-Banach Theorem)
Assume that s is a semi-norm on a vector space V, and assume that φ_0 is a linear functional defined on a subspace $D(\varphi_0) \subset V$, and

$$|\varphi_0(y)| \leq s(y), \quad y \in D(\varphi_0). \tag{3.2}$$

Then there is a linear functional φ on V such that

$$|\varphi(x)| \leq s(x), \quad x \in V, \tag{3.3}$$

and

$$\varphi(y) = \varphi_0(y) \tag{3.4}$$

for $y \in D(\varphi_0)$.

THEOREM 3.7 (Principle of Uniform Boundedness)
Suppose that $\{T_\lambda \mid \lambda \in R\}$ is a family of bounded linear operators from a Banach space V into a normed space W and

$$\sup\{\|T_\lambda x\| \mid \lambda \in R\} < \infty \tag{3.5}$$

Chapter 4

Hilbert Spaces.

One very basic fact about the vector space R^k is that it is possible to define the *angle* between two vectors, giving us the fundamental concept of two vectors being *orthogonal* to each other. This gives the space a rich *geometric* structure that is inherited by a special class of Banach spaces, namely the *Hilbert spaces* to be defined in the following.

4.1 Inner Product Spaces

The first ingredient to be introduced is the *inner product*, the generalization of the scalar product from linear algebra.

DEFINITION 4.1 *Let V be a vector space. An* inner product *is a mapping $(\cdot, \cdot)$ from $V \times V$ into C satisfying*
(i): $(x, y) = \overline{(y, x)}$ for $x, y \in V$,
(ii): $(\alpha x_1 + \beta x_2, y) = \alpha(x_1, y) + \beta(x_2, y)$ for $x_1, x_2, y \in V$ and $\alpha, \beta \in C$,
(iii): $(x, x) \geq 0$ and $(x, x) = 0$ if and only if $x = 0$.

Notice that the conditions imply that $(x, \alpha y_1 + \beta y_2) = \overline{\alpha}(x, y_1) + \overline{\beta}(x, y_2)$, so the inner product is linear in the first argument and conjugate linear in the second.

Example 4.1
In $C_0(R)$ we can define an inner product by

$$(f, g) = \int_R f(x)\overline{g(x)}m(x)dx, \qquad (4.1)$$

where $m(x)$ is a positive, continuous function.

In $L^2(R)$ we have the inner product

$$(f,g) = \int_R f(x)\overline{g(x)}dx, \tag{4.2}$$

which is well defined by Cauchy-Schwarz' inequality.   ⬜

We notice from Example 4.1 that in $L^2(R)$ the norm is given by

$$\|f\|_2 = (f,f)^{\frac{1}{2}}, \tag{4.3}$$

and we will always assume that $L^2(R)$ is equipped with that particular inner product, unless otherwise specified. This is an example of a normed space, where the norm is given by an inner product.

In C^k we can define an inner product by

$$(x,y) = \sum_{i=1}^{k} x_i\overline{y_i}, \tag{4.4}$$

notice how the usual Euclidean norm in C^k is given by this inner product.

It is no coincidence that the inner product induces a norm; it is always true that a space equipped with an inner product is a normed space.

DEFINITION 4.2   *Let V be a vector space and assume that $(\cdot,\cdot)$ is an inner product on V. The induced norm on V is defined by*

$$\|x\| = (x,x)^{\frac{1}{2}}. \tag{4.5}$$

We have not shown that this defines a norm in general, but the only thing that is not trivial is the triangle inequality. But this follows from the following version of Cauchy-Schwarz' inequality:

PROPOSITION 4.1
Let V be a vector space and assume that $(\cdot,\cdot)$ is an inner product on V.
Then

$$|(x,y)|^2 \le (x,x)(y,y) \quad \text{for all} \quad x,y \in V. \tag{4.6}$$

PROOF   If $(y,y) = 0$, we have that $y = 0$, and the inequality is obvious. If $(y,y) \ne 0$, we have for any $\alpha \in C$ that

$$0 \le (x - \alpha y, x - \alpha y) = (x,x) - \alpha(y,x) - \overline{\alpha}(x,y) + \alpha\overline{\alpha}(y,y),$$

and if we take

$$\alpha = \frac{(x,y)}{(y,y)}$$

the result follows. ∎

Now the triangle inequality for the induced norm follows, since we have

$$\begin{aligned}
\|x+y\|^2 &= \|x\|^2 + \|y\|^2 + (x,y) + (y,x) \\
&\leq \|x\|^2 + \|y\|^2 + 2|(x,y)| \\
&\leq \|x\|^2 + \|y\|^2 + 2\|x\|\|y\| \\
&= (\|x\| + \|y\|)^2 .
\end{aligned}$$

Then we can write the Cauchy-Schwarz inequality as

$$|(x,y)| \leq \|x\|\|y\|, \tag{4.7}$$

which is the version we met earlier.

Notice also how the calculations in the proof show that the $=$ in the inequality is valid only if x and y are linearly dependent; that is, $x = \alpha y$ or $y = 0$.

As a simple application of Cauchy-Schwarz' inequality we will show that the inner product is continuous.

PROPOSITION 4.2

Let $(\cdot,\cdot)$ be an inner product on the vector space V, and let V be normed by the induced norm. If $x_n \to x$ and $y_n \to y$, then

$$(x_n, y_n) \to (x,y).$$

PROOF   Since $\|x_n - x\| \to 0$ and $\|y_n - y\| \to 0$, we have that

$$\begin{aligned}
|(x_n,y_n) - (x,y)| &= |(x_n - x, y_n - y) + (x_n - x, y) + (x, y_n - y)| \\
&\leq |(x_n - x, y_n - y)| + |(x_n - x, y)| + |(x, y_n - y)| \\
&\leq \|x_n - x\|\|y_n - y\| + \|x_n - x\|\|y\| + \|x\|\|y_n - y\| \to 0.
\end{aligned}$$

∎

In the special case where the inner product (x,y) is a real number for all x and y, we speak of a *real inner product*. Notice that in a real inner product space, we can define the *angle* between two vectors by

$$\cos(\theta) = \frac{(x,y)}{\|x\|\|y\|}, \tag{4.8}$$

since the right hand side lies between -1 and 1. The most useful notion for our purposes is that of a right angle, implying that $(x,y) = 0$. This leads to the following definition, which applies to both real and complex inner product spaces.

DEFINITION 4.3   *Two vectors x, y in an inner product space are said to be* orthogonal *if $(x, y) = 0$. We will write $x \perp y$ if x and y are orthogonal.*

This gives us an abstract version of Pythagoras' theorem from elementary geometry.

PROPOSITION 4.3
If $x \perp y$, then $\|x + y\|^2 = \|x\|^2 + \|y\|^2$.

PROOF

$$
\begin{aligned}
\|x + y\|^2 &= (x + y, x + y) \\
&= (x, x) + (x, y) + (y, x) + (y, y) \\
&= (x, x) + (y, y) \\
&= \|x\|^2 + \|y\|^2,
\end{aligned}
$$

since $(x, y) = 0$.   ∎

4.2   Hilbert Spaces

We see that inner product spaces have many properties in common with the Euclidean vector spaces we have met in linear algebra, a fact that we will elaborate on in the following. The Euclidean vector spaces are complete spaces; if this is also the case for an inner product space it is a Banach space, and it is called a *Hilbert space*.

DEFINITION 4.4   *A vector space with an inner product that is a Banach space with respect to the induced norm is called a* Hilbert space.

We see that the vector spaces C^k are Hilbert spaces; these are finite dimensional Hilbert spaces. For examples of infinite dimensional Hilbert spaces, take $L^2(R)$, or l^2 with the corresponding inner product

$$
((x_n), (y_n)) = \sum_{k=1}^{\infty} x_k \overline{y}_k. \tag{4.9}
$$

As we shall see later, this is, in a sense, the prototype of a Hilbert space.

Now we will discuss the concept of a basis in the Hilbert space case. In Chapter 2 we defined what we will call a basis in a normed space, and here we emphasize that we only consider separable spaces. There are of course examples of nonseparable Hilbert spaces, but they are rare in applications.

DEFINITION 4.5   *A set of vectors $\{x_k\}$ in an inner product space is called an* orthogonal *set if $(x_i, x_j) = 0$ for $i \neq j$, and $x_j \neq 0$ for all j. If also $(x_j, x_j) = 1$, that is, if all vectors are unit vectors, the set is called* orthonormal.

If an orthonormal set is a basis, we will call the set an orthonormal basis.

We notice that any finite orthogonal set is necessarily linearly independent. Since orthogonal sets (and othonormal sets in particular) are convenient to use as a basis, it is nice to have an algorithm to orthonormalize a given set of linearly independent vectors. The next theorem tells us how.

THEOREM 4.4 (Gram-Schmidt)

Let (y_k) be a sequence of linearly independent vectors in a Hilbert space H. There is an orthonormal sequence (x_k) in H, such that for any $n \in N$ we have

$$span\{x_k\}_{k=1}^n = span\{y_k\}_{k=1}^n.$$

PROOF   Let

$$x_1 = \frac{y_1}{\|y_1\|}, \tag{4.10}$$

$$x_2 = \frac{y_2 - (y_2, x_1)x_1}{\|y_2 - (y_2, x_1)x_1\|}, \tag{4.11}$$

and if we assume that $x_1, x_2, ..., x_j$ are defined, let

$$x_{j+1} = \frac{y_{j+1} - \sum_{k=1}^j (y_{j+1}, x_k)x_k}{\|y_{j+1} - \sum_{k=1}^j (y_{j+1}, x_k)x_k\|}.$$

The sequence (x_k) is orthonormal and satisfies the claim .   ∎

We will now proceed with the process of expanding a vector in a Hilbert space as a series of basis vectors. This is in fact a generalization of the Fourier expansion of a function, as we shall soon see.

PROPOSITION 4.5

Let (x_k) be an orthonormal sequence in the Hilbert space H, and let (α_k) be a sequence of real or complex numbers. The series $\sum_{k=1}^\infty \alpha_k x_k$ is convergent *in H if and only if $\sum_{k=1}^{\infty} |\alpha_k|^2 < \infty$. If this is the case, we have that*

$$\| \sum_{k=1}^{\infty} \alpha_k x_k \|^2 = \sum_{k=1}^{\infty} |\alpha_k|^2. \tag{4.12}$$

PROOF   Let $s_n = \sum_{k=1}^{n} \alpha_k x_k$, and let us show that (s_n) is a Cauchy sequence if and only if $\sum_{k=1}^{\infty} |\alpha_k|^2 < \infty$. We have from Pythagoras' theorem that

$$\| s_{n+p} - s_n \|^2 = \sum_{k=n+1}^{n+p} |\alpha_k|^2,$$

for all n, p in N. This equality shows that (s_n) is a Cauchy sequence if and only if $\sum_{k=1}^{\infty} |\alpha_k|^2$ is convergent.

Moreover, we have that

$$\| \sum_{k=1}^{\infty} \alpha_k x_k \|^2 = \lim_n \| s_n \|^2 = \sum_{k=1}^{\infty} |\alpha_k|^2. \tag{4.13}$$

∎

Notice how this proposition implies that *if a vector $x \in H$ can be expressed as a series $x = \sum_{k=1}^{\infty} \alpha_k x_k$ where (x_k) is an orthonormal sequence, then necessarily the sequence (α_k) is in l^2*, the space of square summable sequences.

From linear algebra we know how to expand a vector $x \in R^k$ in terms of an orthonormal basis $e_1, e_2, ..., e_k$ for R^k, namely $x = \sum_{j=1}^{k} \alpha_j e_j$, where the α_j are uniquely determined by $\alpha_j = (x, e_j)$. We will show that this is also the case in the much more general Hilbert space setting.

THEOREM 4.6 (Bessel's equation and inequality)
Let (x_k) be an orthonormal sequence in the Hilbert space H. Then, for all $x \in H$ and $n \in N$ we have:

$$\| x - \sum_{k=1}^{n} (x, x_k) x_k \|^2 = \| x \|^2 - \sum_{k=1}^{n} |(x, x_k)|^2, \tag{4.14}$$

and

$$\sum_{k=1}^{n} |(x, x_k)|^2 \leq \| x \|^2. \tag{4.15}$$

Moreover,

$$\sum_{k=1}^{\infty} |(x, x_k)|^2 \leq \| x \|^2, \tag{4.16}$$

so

$$\sum_{k=1}^{\infty} (x, x_k) x_k \qquad (4.17)$$

is convergent in H.

PROOF   Let $\alpha_1, \alpha_2, ..., \alpha_n$ be a set of real or complex numbers. Then

$$\|x - \sum_{k=1}^{n} \alpha_k x_k\|^2 = \|x\|^2 - \sum_{k=1}^{n} |(x, x_k)|^2 + \sum_{k=1}^{n} |(x, x_k) - \alpha_k|^2.$$

If we choose $\alpha_k = (x, x_k)$, the first part of the statement follows, and the last part follows from the preceeding proposition.   ∎

Notice that from the proof above we have what is sometimes referred to as the *best approximation theorem*:

COROLLARY 4.7

Let (x_k) be an orthonormal sequence in the Hilbert space H and $\alpha_1, \alpha_2, ..., \alpha_n$ a set of real or complex numbers. Then

$$\|x - \sum_{k=1}^{n} \alpha_k x_k\|^2 \geq \|x - \sum_{k=1}^{n} (x, x_k) x_k\|^2. \qquad (4.18)$$

Now, if we assume that the orthonormal set of vectors we have been considering so far is also a basis for the Hilbert space H, it follows that every vector $x \in H$ has an expansion

$$x = \sum_{k=1}^{\infty} \alpha_k e_k, \qquad (4.19)$$

where (e_k) is the orthonormal sequence of basis vectors, and

$$\alpha_j = (x, e_j) = (\sum_{k=1}^{\infty} \alpha_k e_k, e_j) \qquad (4.20)$$

are the coefficients in the orthonormal expansion. Notice how nicely this fits into our linear algebra terminology. The expansion $x = \sum_{k=1}^{\infty} (x, e_k) e_k$ is usually called the *Fourier expansion* or the *Fourier series* for x with respect to the orthonormal basis (e_k), and the numbers (x, e_j) are called the *Fourier coefficients* for x with respect to the orthonormal basis (e_k).

Now we will address the problem of how to decide if an orthonormal set is "large" enough to be a basis. First we show that it is sufficient for a basis to span a dense subspace.

PROPOSITION 4.8
Let V be a dense subset of the Hilbert space H, and assume that (e_n) is an orthonormal basis for V. Then (e_n) is also an orthonormal basis for H.

PROOF   Let $x \in H$ and $\epsilon > 0$ be given. Take $v \in V$ such that $\|x - v\| < \frac{\epsilon}{2}$ and write $v = \sum_{k=1}^{\infty}(v, e_k)e_k$. There is an $n_0 \in N$ such that for $n > n_0$

$$\left\|v - \sum_{k=1}^{n}(v, e_k)e_k\right\| < \frac{\epsilon}{2},$$

hence

$$\left\|x - \sum_{k=1}^{n}(v, e_k)e_k\right\| < \|x - v\| + \left\|v - \sum_{k=1}^{n}(v, e_k)e_k\right\| < \epsilon,$$

so by the best approximation theorem

$$\left\|x - \sum_{k=1}^{n}(x, e_k)e_k\right\| \le \left\|x - \sum_{k=1}^{n}(v, e_k)e_k\right\| < \epsilon,$$

showing that $x = \lim_n \sum_{k=1}^{n}(x, e_k)e_k$. Hence (e_n) is an orthonormal basis for H.   ∎

We have the following characterization of an orthonormal basis:

PROPOSITION 4.9
An orthonormal sequence (x_k) in a Hilbert space H is an orthonormal basis if and only if one of the following conditions hold:

$$(i): \quad (x, y) = \sum_{k=1}^{\infty}(x, x_k)(x_k, y) \quad \text{for all } x, y \text{ in } H.$$

$$(ii): \quad \|x\|^2 = \sum_{k=1}^{\infty}|(x, x_k)|^2 \quad \text{for all } x \text{ in } H.$$

$$(iii): \quad (x, x_k) = 0 \quad \text{for all } k \text{ implies that } x = 0.$$

PROOF   If $x = \sum_{k=1}^{\infty}(x, x_k)x_k$ and $y = \sum_{k=1}^{\infty}(y, x_k)x_k$, we have immediately (i), and (ii) is a special case of (i). If $(x, x_k) = 0$ for all k, we see from (ii) that $\|x\| = 0$, so $x = 0$, which is condition (iii). On the other hand, assume that (iii) is valid and put $y = \sum_{k=1}^{\infty}(x, x_k)x_k$. Then $(x - y, x_k) = 0$ for all k so $x = y$, and we see that (x_k) is an orthonormal basis. ∎

REMARK 4.1   The equation in (ii) is known as *Parsevals equation*. Notice how this is exactly an infinite dimensional version of the fact from elementary linear algebra, that the square of the length of a vector is the sum of the squares of its coordinates. ∎

The next theorem states that, in a sense, all (separable) Hilbert spaces are alike, expressed by the fact that they are *isometric isomorphic*. That two normed vector spaces are isometric isomorphic means that there exist bijective, linear, and isometric operators mapping one to the other.

THEOREM 4.10 (Riesz-Fischer)
Let H be a Hilbert space. Then H is isometric isomorphic to l^2.

PROOF   Let (e_n) be an orthonormal basis for H. For $x \in H$ we define $Tx \in l^2$ by

$$Tx = ((x, e_n)).$$

It is obvious that T is linear, and from Parsevals equation we have that

$$\|Tx\|^2 = \sum_{k=1}^{\infty} |(x, e_k)|^2 = \|x\|^2,$$

so T is isometric, hence injective. On the other hand, let $(\alpha_n) \in l^2$. Then $\sum_{k=1}^{\infty} \alpha_k e_k$ is convergent in H, and

$$T(\sum_{k=1}^{\infty} \alpha_k e_k) = ((\sum_{k=1}^{\infty} \alpha_k e_k, e_n)) = (\alpha_n),$$

showing that T is also surjective. ∎

We know from linear algebra that any finite dimensional vector space is isomorphic to C^k, say; this is just the fact that we can express a vector by its coordinates. The Riesz-Fischer Theorem is just the infinite dimensional version of this fact. The remarkable news is that the coordinates are now sequences tending rapidly enough to zero in order to be square summable.

4.3   Construction of Hilbert Spaces

After the discussion of the characterization of an orthonormal basis for
a general Hilbert space, we will now look at some examples of how to
construct one.

Example 4.2
Let us first consider the Hilbert space l^2. An orthonormal basis is (e_n),
where $e_1 = (1, 0, 0, ...)$, $e_2 = (0, 1, 0, 0, ...)$, and $e_k = (0, 0, ..., 0, 1, 0, ...)$, the
$'1'$ appearing on the k-th place. $\square$

Example 4.3
Then let us consider the *real $L^2(I)$*, where I is a closed, bounded interval.
We know that the set of polynomials, $P(I)$, is dense, and since the sequence
$1, t, t^2, ..., t^k, ...$, that is, $(t^n), n \geq 0$, obviously is a basis for $P(I)$, it is also
a basis for $L^2(I)$. But it is not an orthonormal basis, so we will apply the
Gram-Schmidt orthonormalization procedure to it. We will consider the
case where $I = [-1; 1]$. Since we consider the real $L^2(I)$, the inner product
is now

$$(f, g) = \int_{-1}^{1} f(t)g(t)dt, \qquad (4.21)$$

and we see that the polynomials 1 and t are already orthogonal. We normal-
ize them to get $p_0(t) = \frac{1}{\sqrt{2}}$ and $p_1(t) = \sqrt{\frac{3}{2}}t$. Next we find the polynomial

$$t^2 - \left(\int_{-1}^{1} t^2 p_0(t)dt \cdot p_0(t) + \int_{-1}^{1} t^2 p_1(t)dt \cdot p_1(t)\right) = t^2 - \left(\frac{1}{\sqrt{2}}\right)^2 \int_{-1}^{1} t^2 dt = t^2 - \frac{1}{3}$$

orthogonal to p_0 and p_1, with the norm

$$\|t^2 - \frac{1}{3}\|_2^2 = \int_{-1}^{1} |t^2 - \frac{1}{3}|^2 dt = \frac{8}{45}.$$

So when we normalize we get $p_2(t) = \frac{3}{2}\sqrt{\frac{5}{2}}(t^2 - \frac{1}{3})$. Continuing this way, we
will get what is known as the *normalized Legendre polynomials*, which are
usually written in the form $\sqrt{\frac{2n+1}{2}}P_n(t)$, where the $P_n(t)$ are the *Legendre
polynomials*, given by

$$P_n(t) = \frac{1}{2^n n!}\left(\frac{d}{dt}\right)^n ((t^2 - 1)^n).$$

The sequence of normalized Legendre polynomials is an orthonormal basis for $L^2([-1;1])$, and it can be shown that P_n are solutions to the *Legendre differential equation*

$$\frac{d}{dt}((1-t^2)\frac{du}{dt}) + n(n+1)u = 0, \quad t \in [-1;1], n \geq 0$$

appearing in quantum mechanics problems and other problems with spherical symmetry. $\square$

Example 4.4
Let us now consider (real) $L^2([0;\pi])$ and find an orthonormal basis consisting of trigonometric functions. It is easy to see that the sequence

$$e_0(t) = \frac{1}{\sqrt{\pi}}, \quad e_1(t) = \sqrt{\frac{2}{\pi}}\cos(t), \quad ..., e_k(t) = \sqrt{\frac{2}{\pi}}\cos(kt), ...$$

is orthonormal. To show that it also spans $L^2([0;\pi])$, we notice that since

$$(\cos(t))^k = (\frac{1}{2}(e^{it} + e^{-it}))^k,$$

any power $(\cos(t))^k$ of $\cos(t)$ can be written as a linear combination of the elements of the sequence (e_n). Let $f \in C([0;\pi])$ and define $g \in C([-1;1])$ by $g(t) = f(\text{Arccos}(t))$. For any $\epsilon > 0$, Weierstrass' approximation theorem provides us with a polynomial $P_n(t) = \sum_{k=0}^{n} a_k t^k$, satisfying

$$|g(t) - P_n(t)| < \epsilon \quad \text{for} \quad t \in [-1;1];$$

hence

$$|f(s) - P_n(\cos(s))| = |f(s) - \sum_{k=0}^{n} a_k (\cos(s))^k| < \epsilon \quad \text{for} \quad s \in [0;\pi].$$

Then

$$\|f - P_n(\cos(\cdot))\|_2 \leq \epsilon\sqrt{\pi},$$

and we can approximate any $f \in C([0;\pi])$ with a linear combination of elements from the sequence. By the density of $C([0;\pi])$ in $L^2([0;\pi])$ we have thus shown that (e_n) is an orthonormal basis. But what about a sequence of sines? It is easy to see that the sequence

$$e_1(t) = \sqrt{\frac{2}{\pi}}\sin(t), \quad e_2(t) = \sqrt{\frac{2}{\pi}}\sin(2t), \quad ..., e_k(t) = \sqrt{\frac{2}{\pi}}\sin(kt), ...$$

is orthonormal. In order to show that it is also a basis, we will show that the sequence is total. So let $f \in L^2([0;\pi])$ and assume that

$$(f(t), \sin(kt)) = \int_0^{\pi} f(t)\sin(kt)dt = 0 \quad \text{for} \quad k \geq 1.$$

Then, for $k \geq 0$ we find

$$
\begin{aligned}
(f(t)\sin(t), \cos(kt)) &= \int_0^\pi f(t)\sin(t)\cos(kt)dt \\
&= \frac{1}{2}\int_0^\pi f(t)\sin((k+1)t)dt - \frac{1}{2}\int_0^\pi f(t)\sin((k-1)t)dt \\
&= \frac{1}{2}((f(t),\sin((k+1)t)) - (f(t),\sin((k-1)(t)))) \\
&= 0,
\end{aligned}
$$

and since the cosines form an orthonormal basis, this implies that $f(t)\sin(t)$ is the zero function in $L^2([0;\pi])$, so $f = 0$ in $L^2([0;\pi])$. So the sequence of sines is an orthonormal basis.   ⬚

Notice how the well-known Fourier cosine and sine series for a well-behaved function f on $[0;\pi]$ appear as

$$
f = \sum_k (f, e_k)e_k, \tag{4.22}
$$

and we see why there are severe problems in the elementary analysis of Fourier series. First the function f must belong to $L^2([0;\pi])$; next we must understand that the "=" above is in L^2-sense, meaning that

$$
\lim_n \left\| f - \sum_{k=1}^n (f, e_k)e_k \right\|_2 = 0, \tag{4.23}
$$

(in the sine case). There is no immediate translation of this into the more conceptually clear forms of convergence (pointwise/uniform) we have met in elementary analysis. A Fourier series is the "coordinate expansion" of a vector in $L^2(I)$, and we must be extremely careful if we want to interpret it as something else.

Example 4.5
Let us prove that the classical Fourier series for a C^1-function converges pointwise.

We will consider $f \in C^1([-\pi;\pi])$, and we choose to expand f in an orthonormal basis for $L^2([-\pi;\pi])$ consisting of cosine and sine functions.

The series is then

$$
f(x) = \frac{1}{2}A_0 + \sum_{n=1}^\infty (A_n \frac{1}{\sqrt{\pi}}\cos(nx) + B_n \frac{1}{\sqrt{\pi}}\sin(nx)),
$$

with the coefficients

$$A_n = \frac{1}{\sqrt{\pi}} \int_{-\pi}^{\pi} f(y) \cos(ny) dy, \quad n = 0, 1, 2, \ldots$$

$$B_n = \frac{1}{\sqrt{\pi}} \int_{-\pi}^{\pi} f(y) \sin(ny) dy, \quad n = 1, 2, 3, \ldots$$

and we know that the series is convergent in $L^2([-\pi; \pi])$, which is the meaning of the $=$ above. But let us show that in this case, where $f \in C^1$, it is also an ordinary equality.

The partial sum is

$$s_n(x) = \frac{1}{2} A_0 + \sum_{k=1}^{n} (A_k \frac{1}{\sqrt{\pi}} \cos(kx) + B_k \frac{1}{\sqrt{\pi}} \sin(kx)),$$

and by plugging in the expressions for the coefficients we get

$$s_n(x) = \frac{1}{2\pi} \int_{-\pi}^{\pi} (1 + 2 \sum_{k=1}^{n} (\cos(ky) \cos(kx) + \sin(ky) \sin(kx))) f(y) dy.$$

The expression in the summation can be written as the cosine of a difference of angles, so if we introduce the *Dirichlet kernel*:

$$D_n(\theta) = 1 + 2 \sum_{k=1}^{n} \cos(k\theta), \tag{4.24}$$

we can write the partial sum as

$$s_n(x) = \frac{1}{2\pi} \int_{-\pi}^{\pi} D_n(x - y) f(y) dy.$$

Notice that the Dirichlet kernel has period 2π and satisfies

$$\frac{1}{2\pi} \int_{-\pi}^{\pi} D_n(\theta) d\theta = 1.$$

Moreover, by using the complex exponentials, we see that

$$D_n(\theta) = 1 + \sum_{k=1}^{n} (e^{ik\theta} + e^{-ik\theta})$$

$$= \sum_{k=-n}^{n} (e^{ik\theta} + e^{-ik\theta})$$

$$= \frac{e^{-in\theta} - e^{i(n+1)\theta}}{1 - e^{i\theta}}$$

$$= \frac{\sin((n + \frac{1}{2})\theta)}{\sin(\frac{\theta}{2})}.$$

Since D_n is obviously an even function, we can substitute $\theta = y - x$ in the formula for $s_n(x)$, and since we can consider both D_n and f periodically extended to a larger interval, we can write

$$s_n(x) = \frac{1}{2\pi} \int_{-\pi}^{\pi} D_n(\theta) f(x + \theta) d\theta.$$

Now, subtraction of $f(x) = f(x) \cdot 1$ gives

$$s_n(x) - f(x) = \frac{1}{2\pi} \int_{-\pi}^{\pi} D_n(\theta)(f(x + \theta) - f(x)) d\theta$$

$$= \frac{1}{2\pi} \int_{-\pi}^{\pi} h(\theta) \sin((n + \frac{1}{2})\theta) d\theta,$$

where we have introduced the function (recall that x is fixed):

$$h(\theta) = \frac{f(x + \theta) - f(x)}{\sin(\frac{\theta}{2})}. \tag{4.25}$$

Now notice that the functions $\varphi_k(\theta) = \sin((k + \frac{1}{2})\theta)$ form an orthogonal set in $L^2([-\pi; \pi])$, so by Bessel's inequality

$$s_n(x) - f(x) = \frac{1}{2\pi} \int_{-\pi}^{\pi} h(\theta) \sin((n + \frac{1}{2})\theta) d\theta = \frac{1}{2\pi}(h, \varphi_n)$$

converges to zero, *provided that* h *belongs to* $L^2([-\pi; \pi])$, which is the only thing now to show. Since h is a fraction of continuous functions, we have only to show that

$$\int_{-\pi}^{\pi} (\frac{f(x + \theta) - f(x)}{\sin(\frac{\theta}{2})})^2 d\theta$$

is finite. The only possible problem is where $\theta = 0$, but at that point, by l'Hospitals rule (since $f \in C^1$) :

$$\lim_{\theta \to 0} \frac{f(x + \theta) - f(x)}{\sin(\frac{\theta}{2})} = \lim_{\theta \to 0} \frac{f(x + \theta) - f(x)}{\frac{\theta}{2}} \cdot \frac{\frac{\theta}{2}}{\sin(\frac{\theta}{2})} = 2f'(x).$$

Hence the integral is finite, so h belongs to $L^2([-\pi; \pi])$ and the partial sum converges to $f(x)$.

$\square$

Example 4.6
In the Hilbert space $L^2(R)$ we can construct an orthonormal basis from the linearly independent sequence

$$e^{-\frac{t^2}{2}}, \quad te^{-\frac{t^2}{2}}, \quad t^2 e^{-\frac{t^2}{2}}, \dots$$

by applying the Gram-Schmidt orthonormalization procedure. This gives us an orthonormal sequence of the form

$$e_k(t) = \frac{(-1)^k e^{\frac{t^2}{2}}}{\sqrt{2^k k! \sqrt{\pi}}} \frac{d^k}{dt^k}(e^{-t^2}), \quad k \geq 0, \tag{4.26}$$

and with some effort it can be shown to be an orthonormal basis. It is convenient to write this as

$$e_k(t) = \frac{e^{-\frac{t^2}{2}}}{\sqrt{2^k k! \sqrt{\pi}}} H_k(t), \quad k \geq 0, \tag{4.27}$$

where the $H_k(t)$ are the *Hermite polynomials*:

$$H_k(t) = (-1)^k e^{t^2} \frac{d^k}{dt^k}(e^{-t^2}), \quad k \geq 0. \tag{4.28}$$

The first Hermite polynomials are

$$H_0(t) = 1, \quad H_1(t) = 2t, \quad H_2(t) = 4t^2 - 2, \quad H_3(t) = 8t^3 - 12t, \ldots$$

and it can be shown that H_k is the solution to the *Hermite differential equation*

$$\frac{d^2 u}{dt^2} - 2t \frac{du}{dt} + 2ku = 0, \quad \text{for} \quad t \in R \tag{4.29}$$

frequently appearing in applications. ▯

Example 4.7
If we instead consider the Hilbert space $L^2([0; \infty[)$, and start out with the linearly independent sequence

$$e^{-\frac{t}{2}}, \quad te^{-\frac{t}{2}}, \quad t^2 e^{-\frac{t}{2}}, \ldots$$

the Gram-Schmidt orthonormalization procedure gives us an orthonormal basis of the form

$$e_k(t) = e^{-\frac{t}{2}} L_k(t), \quad k \geq 0, \tag{4.30}$$

where the L_k are the *Laguerre polynomials*

$$L_k(t) = \frac{1}{k!} e^t \frac{d^k}{dt^k}(t^k e^{-t}) = \sum_{j=0}^{k} \frac{(-1)^j}{j!} \binom{k}{j} t^j, \tag{4.31}$$

solving the *Laguerre differential equation*

$$t\frac{d^2u}{dt^2} + (1-t)\frac{du}{dt} + ku = 0, \quad t > 0 \tag{4.32}$$

which is also very common in applications. The first Laguerre polynomials
are

$$L_0(t) = 1$$
$$L_1(t) = 1 - t$$
$$L_2(t) = 1 - 2t + \frac{1}{2}t^2$$
$$L_3(t) = 1 - 3t + \frac{3}{2}t^2 - \frac{1}{6}t^3.$$

☐

Now it is natural to pose the question: Why do we care about *orthonormal* expansions when we have a simple, linearly independent set that is dense? In $L^2([-1;1])$, the sequence $1, t, t^2, \dots$ is much simpler than the Legendre polynomials; they span the whole space, so why do we not just use them? There are many reasons, but the most important is perhaps the following:

Let (e_n) be an orthonormal basis in a Hilbert space H, and let for each $n \in N$, E_n be the finite dimensional subspace spanned by $e_1, e_2, \dots, e_n$. For any $x \in H$, the vector $s_n = \sum_{k=1}^{n}(x, e_k)e_k$ is then the vector from E_n that is the best approximation to x. We see that we can calculate the best approximating vector from E_{n+1} in the very simple way $s_{n+1} = s_n + (x, e_{n+1})e_{n+1}$. If we consider a basis (f_n) for H that is *not* orthogonal, then we can consider the finite dimensional subspace F_n spanned by the vectors $f_1, f_2, \dots, f_n$. The vector from F_n that is closest to $x \in H$ can again be expressed as $s_n = \sum_{k=1}^{n} \alpha_k f_k$, where the α_k's are determined uniquely by

$$0 = (x - \sum_{k=1}^{n} \alpha_k f_k, f_j), \quad j = 1, 2, \dots, n.$$

The problem is that the α_k's in general *depend on n*, so if we want the vector from F_{n+1} that is closest to x, we must calculate *all* the $n + 1$ coordinates again.

Example 4.8
Let $f \in L^2([-\pi; \pi])$. By the *complex Fourier series* for f we understand the series

$$\sum_{k \in Z} c_k e^{ikx}, \tag{4.33}$$

where the coefficients c_k are calculated as

$$c_k = \frac{1}{2\pi} \int_{-\pi}^{\pi} f(t)e^{-ikt}\,dt. \tag{4.34}$$

This is just the expansion of f with respect to the orthonormal basis (e_n), where

$$e_k(t) = \frac{1}{\sqrt{2\pi}}e^{ikt}, \quad k \in Z. \tag{4.35}$$

□

4.4 Orthogonal Projection and Complement

We noticed in the previous section that if (e_n) is an orthonormal basis for the Hilbert space H, then $s_n = \sum_{k=1}^{n}(x, e_k)e_k$ is the vector from E_n, the closed subspace spanned by the n first basis vectors that are closest to x. We call s_n the *orthogonal projection* of x onto E_n. Notice that for fixed x, $y = s_n$ is the unique solution to the problem of minimizing $\|x - y\|$ for $y \in E_n$. We will take a closer look at this problem in a sligthly more general frame. Recall that a set K in a vector space is *convex* if $\lambda x + (1 - \lambda)y$ is in K for $0 < \lambda < 1$ whenever x and y are in K. We need the following lemma, usually called the *parallelogram law:*

LEMMA 4.11
In a vector space where the norm is induced by an inner product, we have that

$$\|x + y\|^2 + \|x - y\|^2 = 2(\|x\|^2 + \|y\|^2) \tag{4.36}$$

for all x and y.

PROOF   Just calculate the left hand side.   ∎

PROPOSITION 4.12
Let K be a closed convex subset of the Hilbert space H. For any $x_0 \in H$, there is a uniquely determined $y_0 \in K$, such that

$$\|x_0 - y_0\| \le \|x_0 - y\| \tag{4.37}$$

for all $y \in K$.

PROOF   Let $\delta = \inf\{\|x_0 - y\| \mid y \in K\}$; then there is a sequence (y_k) in K such that $\|x_0 - y_k\| \to \delta$. We will show that (y_k) is a Cauchy sequence. From the parallelogram law we have that

$$
\begin{aligned}
\|y_n - y_m\|^2 &= \|(y_n - x_0) + (x_0 - y_m)\|^2 \\
&= 2\|y_n - x_0\|^2 + 2\|x_0 - y_m\|^2 - \|y_n + y_m - 2x_0\|^2 \\
&= 2\|y_n - x_0\|^2 + 2\|x_0 - y_m\|^2 - 4\|\frac{1}{2}(y_n + y_m) - x_0\|^2 \\
&\leq 2\|y_n - x_0\|^2 + 2\|x_0 - y_m\|^2 - 4\delta^2.
\end{aligned}
$$

The inequality follows from the fact that $\frac{1}{2}(y_n + y_m)$ is in K due to the convexity assumption. The last expression converges to 0 for $m, n \to \infty$, so (y_k) is a Cauchy sequence, hence convergent, and since K is closed, the limit point y_0 belongs to K and $\delta = \|x_0 - y_0\|$.

Assume that also $\delta = \|x_0 - z_0\|$ for some $z_0 \in K$. Then

$$
\begin{aligned}
\|y_0 - z_0\|^2 &= \|(y_0 - x_0) + (x_0 - z_0)\|^2 \\
&= 2\|y_0 - x_0\|^2 + 2\|x_0 - z_0\|^2 - 4\|\frac{1}{2}(y_0 + z_0) - x_0\|^2 \\
&\leq 2\delta^2 + 2\delta^2 - 4\delta^2 = 0,
\end{aligned}
$$

so $y_0 = z_0$.   ∎

In the special case where K is a closed subspace, the minimizer y_0 is just the orthogonal projection from above; this proposition shows, however, that we can define an orthogonal projection onto any closed, convex set in H.

DEFINITION 4.6   *Let M and N be nonempty subsets of a Hilbert space H. We say that M and N are* orthogonal *and write $M \perp N$ if $(x, y) = 0$ for all $x \in M$ and $y \in N$.*

For a nonempty subset M of H, we define the orthogonal complement *to M, denoted $M^\perp$, by*

$$
M^\perp = \{y \in H \mid (x, y) = 0 \quad \text{for all } x \in M\}. \tag{4.38}
$$

It is easy to show that $M^\perp$ is a closed subspace of H. Notice also the connection to the orthonormal basis concept: if (e_n) is an orthonormal basis for H, we can take a subset $I \subset N$ and define M as the closed subspace of H spanned by the set $\{e_k \mid k \in I\}$. (Any closed subspace of a separable Hilbert space can be defined in this way.) Then it is easy to see that

$$
M^\perp = \{x \in H \mid (x, e_k) = 0 \quad \text{for all } k \in I\}. \tag{4.39}
$$

DEFINITION 4.7   *Let M and N be closed subspaces of the Hilbert space H, with $M \perp N$. We define the orthogonal sum of M and N, denoted $M \oplus N$, by*

$$M \oplus N = \{z \in H \mid z = x + y, \quad x \in M, \quad y \in N\}. \qquad (4.40)$$

It is obvious that the representation $z = x + y$ is unique and it is also clear that $M \oplus N$ is a subspace of H. That it is a *closed* subspace is shown in the following way: if $z_n = x_n + y_n$ is a convergent sequence from $M \oplus N$ with $z_n \to z_0$, we must show that $z_0 \in M \oplus N$. But since

$$\|z_n - z_m\|^2 = \|x_n - x_m\|^2 + \|y_n - y_m\|^2, \qquad (4.41)$$

we see that (x_n) and (y_n) are both Cauchy sequences, hence convergent with $x_n \to x_0 \in M$ and $y_n \to y_0 \in N$ since M and N are closed. So $z_0 = x_0 + y_0 \in M \oplus N$, and $M \oplus N$ is closed.

We are now able to formulate the next theorem, which is known as the *projection theorem*. It is the generalization of the fact from finite dimensional linear algebra, e.g., $R^n = R^{n-k} \oplus R^k$, for $0 \le k < n$; here we identify R^{n-k} with the subspace of R^n where the last k coordinates are zero, and R^k is the one with zeroes on the first $n - k$ coordinates. In infinite dimension the analogue is far from trivial.

THEOREM 4.13
If M is a closed subspace of the Hilbert space H, then $H = M \oplus M^{\perp}$.

PROOF   Let $z \in H$. According to the general projection proposition, there is a unique $x \in M$ such that $\|z - x\| \le \|z - x'\|$, for all $x' \in M$. Then we define $y = z - x$, and we have now only to show that $y \in M^{\perp}$.

For $\lambda \in C$ and $x' \in M$ we have that

$$\begin{aligned}
\|y\|^2 &= \|z - x\|^2 \\
&\le \|z - x - \lambda x'\|^2 \\
&= \|y - \lambda x'\|^2 \\
&= \|y\|^2 - \overline{\lambda}(y, x') - \lambda(x', y) + |\lambda|^2 (x', x'),
\end{aligned}$$

and by taking $\lambda = (y, x')$ and $\|x'\|^2 = 1$ we get

$$\|y\|^2 \le \|y\|^2 - |\lambda|^2.$$

This shows that necessarily $0 = \lambda = (y, x')$ for all $x' \in M$, that is, $y \in M^{\perp}$.

∎

We will now proceed with the development of the Hilbert space theory by proving one of the most remarkable facts about these, namely *Riesz'* *representation theorem*, stating that a Hilbert space identifies with its dual. (Recall from Chapter 3, that the dual H^* of a Banach space H is the (Banach) space $B(H, C)$ of bounded, linear functionals on H.)

THEOREM 4.14 (Riesz' representation theorem)
Let φ be a continuous linear functional on a Hilbert space H. Then there is a unique $z \in H$ such that $\varphi(x) = (x, z)$ for all $x \in H$.

PROOF   Let $N = \{x \in H \mid \varphi(x) = 0\}$. Since φ is continuous, N is closed. If $N = H$, we take $z = 0$ to do the job, otherwise we write $H = N \oplus N^\perp$ according to the previous theorem. Take $y_0 \in N^\perp$ with $\|y_0\| = 1$, and consider for any $x \in H$ the vector $y = \varphi(x)y_0 - \varphi(y_0)x$. Since $\varphi(y) = 0$ for all x, y must belong to N, hence

$$0 = (y, y_0) = (\varphi(x)y_0 - \varphi(y_0)x, y_0) = \varphi(x)(y_0, y_0) - \varphi(y_0)(x, y_0) = \varphi(x) - (x, z),$$

where $z = \overline{\varphi(y_0)}y_0$. Therefore $\varphi(x) = (x, z)$.

To show uniqueness, assume that we have two representations $\varphi(x) = (x, v) = (x, w)$. Then take $x = v - w$ and observe that

$$0 = (x, v) - (x, w) = (x, v - w) = \|v - w\|^2$$

implying that $v - w = 0$, and we have uniqueness.   ∎

4.5   Weak Convergence

We will now introduce the concept of *weak convergence* in a Hilbert space. This can also be introduced in the more general setting of Banach spaces, but due to Riesz' representation theorem the situation is much simpler in Hilbert spaces.

DEFINITION 4.8   *A sequence (x_n) in a Hilbert space H is* weakly convergent *with* weak limit *x if, for all $y \in H$, the sequence (x_n, y) converges to (x, y) in the usual sense, and we write $x_n \rightharpoonup x$ in this case.*

Using Riesz' representation theorem, this is equivalent to the statement that $\varphi(x_n)$ converges to $\varphi(x)$ for all bounded linear functionals φ on H.

It is trivial that the weak limit is unique, because if $x_n \rightharpoonup x$ and $x_n \rightharpoonup z$, then $(x, y) = (z, y)$ for all $y \in H$, hence $x - z \in H^\perp = \{0\}$.

Example 4.9
Let (e_n) be an orthonormal sequence in the Hilbert space H. For $x \in H$, we have from Bessels inequality that

$$\sum_{k=1}^{\infty} |(x, e_n)|^2 \leq \|x\|^2,$$

so $(e_n, x) \to 0 = (0, x)$ for all $x \in H$, showing that (e_n) converges weakly to 0. On the other hand,

$$\|e_n - e_m\|^2 = 2$$

for all $n \neq m$, so (e_n) cannot converge in the usual sense. This justifies the name *weak convergence*.   ☐

PROPOSITION 4.15
Let (x_n) be a sequence in H and assumme that $x_n \to x$. Then $x_n \rightharpoonup x$.

PROOF   Assume that $\|x_n - x\| \to 0$. From Cauchy-Schwarz' inequality we find

$$|(x_n, y) - (x, y)| = |(x_n - x, y)| \leq \|x_n - x\| \|y\| \to 0,$$

for all $y \in H$, hence $x_n \rightharpoonup x$.   ■

The difference between convergence and weak convergence is a genuine infinite dimensional phenomenon; if the Hilbert space is finite-dimensional, weak convergence will imply convergence.

PROPOSITION 4.16
Let H be a finite-dimensional Hilbert space. Then any weakly convergent sequence will be convergent.

PROOF   Assume that $dim(H) = k$ and let $e_1, e_2, ..., e_k$ be an orthonormal basis for H. So if (x_n) converges weakly to x, we will write

$$x_n = \alpha_{1n} e_1 + \alpha_{2n} e_2 + ... + \alpha_{kn} e_k$$

and

$$x = \alpha_1 e_1 + \alpha_2 e_2 + ... + \alpha_k e_k,$$

and since $(x_n, y) \to (x, y)$ for all $y \in H$, we can take $y = e_j$, $j = 1, 2, ..., k$ and deduce that $\alpha_{jn} \to \alpha_j$ $j = 1, 2, ..., k$. Then

$$\|x_n - x\|^2 = \|\sum_{j=1}^{k} (\alpha_{jn} - \alpha_j) e_j\|^2 = \sum_{j=1}^{k} |\alpha_{jn} - \alpha_j|^2 \to 0,$$

so $x_n \to x$. ∎

PROPOSITION 4.17
A weakly convergent sequence is bounded.

PROOF   The proof is left as an exercise.   ∎

Chapter 5

Operators on Hilbert Spaces

The Hilbert spaces are Banach spaces, and all the results we have on bounded operators from Chapter 3 apply in the Hilbert spaces, but due to the simple nature of the dual of a Hilbert space that the Riesz' representation theorem has revealed, we can establish a rich operator theory.

As a first application of Riesz' representation theorem, we will define the *adjoint* of a bounded operator on a Hilbert space. In finite dimensions, where the bounded linear operators are represented by matrices, this is just the conjugate transpose. But in infinite dimension, the situation is again much more delicate.

5.1   The Adjoint of a Bounded Operator

THEOREM 5.1
Let H be a Hilbert space and $T \in B(H)$.
  There is a unique operator $T^* \in B(H)$ satisfying

$$(Tx, y) = (x, T^*y) \quad \text{for all} \quad x, y \in H, \tag{5.1}$$

and we have $\|T^*\| = \|T\|$.
  The operator T^* is called the adjoint *of T.*

PROOF   Take a $y \in H$. Since T is linear and bounded, we can define a bounded linear functional on H by

$$\varphi_y(x) = (Tx, y) \tag{5.2}$$

for all $x \in H$. That φ_y is bounded follows from Cauchy-Schwarz' inequality:

$$|\varphi_y(x)| = |(Tx, y)| \leq \|Tx\|\|y\| \leq \|T\|\|x\|\|y\|.$$

From Theorem 4.14 there is then a unique $z \in H$ such that

$$\varphi_y(x) = (x, z) \tag{5.3}$$

for all $x \in H$. Since z depends on y, we have in this way defined an operator $T^*y = z$ satisfying

$$(Tx, y) = (x, T^*y) \quad \text{for all} \quad x, y \in H. \tag{5.4}$$

To show that T^* is linear, we take $y_1, y_2 \in H$ and $\lambda_1, \lambda_2 \in C$, and calculate for all $x \in H$:

$$
\begin{aligned}
(x, T^*(\lambda_1 y_1 + \lambda_2 y_2)) &= (Tx, \lambda_1 y_1 + \lambda_2 y_2) \\
&= (Tx, \lambda_1 y_1) + (Tx, \lambda_2 y_2) \\
&= \overline{\lambda_1}(Tx, y_1) + \overline{\lambda_2}(Tx, y_2) \\
&= \overline{\lambda_1}(x, T^*y_1) + \overline{\lambda_2}(x, T^*y_2) \\
&= (x, \lambda_1 T^*y_1) + (x, \lambda_2 T^*y_2) \\
&= (x, \lambda_1 T^*y_1 + \lambda_2 T^*y_2)
\end{aligned}
$$

from which we deduce that

$$T^*(\lambda_1 y_1 + \lambda_2 y_2) = \lambda_1 T^*y_1 + \lambda_2 T^*y_2.$$

Moreover, for all $y \in H$:

$$
\begin{aligned}
\|T^*y\|^2 &= (T^*y, T^*y) \\
&= (TT^*y, y) \\
&\leq \|TT^*y\|\|y\| \\
&\leq \|T\|\|T^*y\|\|y\|,
\end{aligned}
$$

and we see that $\|T^*y\| \leq \|T\|\|y\|$, since the case $\|T^*y\| = 0$ is trivial. This shows that $\|T^*\| \leq \|T\|$, and T^* is bounded. Substituting T for T^* above gives the inequality $\|T\| \leq \|T^*\|$, so we see that $\|T^*\| = \|T\|$. ∎

REMARK 5.1   Notice that we have that $(T^*)^* = T$ and $(ST)^* = T^*S^*$. Moreover, from the proof above we see that

$$\|T^*T\| = \|T\|^2.$$

∎

The following proposition is well known from linear algebra, but again we must be careful in infinite dimensions.

PROPOSITION 5.2

Let $T \in B(H)$ where H is a Hilbert space. Then

$$H = ker(T^*) \oplus \overline{T(H)}.$$

PROOF Since $ker(T^*)$ is closed, we have only to show that $T(H)^{\perp} = ker(T^*)$, due to the projection theorem. So let $y \in T(H)^{\perp}$, then $0 = (Tx, y) = (x, T^*y)$ for all $x \in H$, hence $T^*y = 0$ and $y \in ker(T^*)$. On the other hand, if $y \in ker(T^*)$, then $0 = (x, T^*y) = (Tx, y)$ for all $x \in H$, so $y \in T(H)^{\perp}$. Hence $T(H)^{\perp} = ker(T^*)$, and the result follows. ∎

Example 5.1

A bounded linear operator $T : C^k \to C^k$ is represented by a $k \times k$-matrix (T_{ij}). For $x = (x_1, x_2, ..., x_k)$ and $y = (y_1, y_2, ..., y_k)$ in C^k we have

$$(Tx, y) = \sum_{i=1}^{k} (Tx)_i \overline{y_i}$$

$$= \sum_{i=1}^{k} (\sum_{j=1}^{k} T_{ij} x_j) \overline{y_i}$$

$$= \sum_{j=1}^{k} x_j \sum_{i=1}^{k} T_{ij} \overline{y_i}$$

$$= \sum_{i=1}^{k} x_i (\overline{\sum_{j=1}^{k} \overline{T_{ji}} y_j})$$

$$= (x, T^*y)$$

so that $(T^*y)_i = \sum_{j=1}^{k} \overline{T_{ji}} y_j)$. Hence we see that T^* is represented by the matrix $(\overline{T_{ji}})$, which is the conjugate transpose of (T_{ij}). ▯

Example 5.2

Let $k \in C(I \times I)$ where I is a closed, bounded interval. We can define a linear operator $K : L^2(I) \to L^2(I)$ by

$$Kf = \int_I k(\cdot, y) f(y) dy.$$

Since for $f \in L^2(I)$:

$$\|Kf\|^2 = \int_I |\int_I k(x,y)f(y)dy|^2 dx$$
$$\leq \int_I (\int_I |k(x,y)f(y)|dy)^2 dx$$
$$\leq \|k\|_\infty^2 \int_I (\int_I 1 \cdot |f(y)|dy)^2 dx$$
$$\leq M\|k\|_\infty^2 \|f\|^2,$$

we see that K is bounded. (Notice that we can use Cauchy-Schwarz' inequality, because the constant function 1 is in $L^2(I)$ for any bounded interval.) To determine the adjoint K^*, we consider $f, g \in L^2(I)$ and calculate

$$(Kf, g) = \int_I Kf(x)\overline{g(x)}dx$$
$$= \int_I (\int_I k(x,y)f(y)dy)\overline{g(x)}dx$$
$$= \int_I f(y)(\int_I k(x,y)\overline{g(x)}dx)dy$$
$$= \int_I f(y)(\overline{\int_I \overline{k(x,y)}g(x)dx})dy,$$

and we see that

$$K^*g(y) = \int_I \overline{k(x,y)}g(x)dx. \tag{5.5}$$

Hence K^* is the integral operator with conjugate transpose kernel $k^*(x,y) = \overline{k(y,x)}$. (We use here the fact that the order of integration can be reversed, this is not a trivial fact but follows from measure theory, the so-called *Fubinis theorem*. The result extends without problems to the case where $k \in L^2(I \times I)$. The operator K can, in this case, be considered as the limit of a sequence of bounded operators (K_n) with corresponding kernels (k_n) chosen such that $\lim_n k_n = k$ in $L^2(I \times I)$, which is possible since $C(I \times I)$ is dense in $L^2(I \times I)$. The class of operators of this form is called the *Hilbert-Schmidt operators*, and we will take a closer look at them later on.)

We see that if the function k is real and symmetric, then $K = K^*$. This is a particularly important case.

□

DEFINITION 5.1   *Let H be a Hilbert space and $T \in B(H)$. If $T = T^*$ we say that T is* self-adjoint.

The self-adjoint operators are very common in applications, and they have many nice properties.

LEMMA 5.3
Let T be a bounded, self-adjoint operator on a Hilbert space H. Then (Tx, x) is real for all $x \in H$.

PROOF
$$(Tx, x) = (x, Tx) = \overline{(Tx, x)}.$$

∎

The next result gives an expression for the norm of a self-adjoint operator that perhaps at first looks far from simple, but notice that the expression can be calculated by maximizing a "quadratic form" (Tx, x) that is linear in T, whereas $\|T\|$ in general is calculated by maximizing $\sqrt{(Tx, Tx)}$, which is quadratic in T.

PROPOSITION 5.4
Let T be a bounded, self-adjoint operator on a Hilbert space H. Then

$$\|T\| = \sup\{|(Tx, x)| \mid x \in H, \quad \|x\| = 1\}. \tag{5.6}$$

PROOF Let $M_T = \sup\{|(Tx, x)| \mid x \in H, \quad \|x\| = 1\}$. From Cauchy-Schwarz' inequality we have for $\|x\| = 1$ that

$$|(Tx, x)| \leq \|Tx\|\|x\| \leq \|T\|\|x\|\|x\| = \|T\|,$$

so $M_T \leq \|T\|$.

On the other hand, a simple calculation shows that

$$(T(x + y), x + y) - (T(x - y), x - y) = 2(Tx, y) + 2(Ty, x),$$

where each term on the left hand side is real since T is self-adjoint. So from the definition of M_T we have that

$$(T(x + y), x + y) \leq M_T \|x + y\|^2,$$
$$-(T(x - y), x - y) \leq M_T \|x - y\|^2.$$

Since from the parallelogram law

$$\|x + y\|^2 + \|x - y\|^2 = 2(\|x\|^2 + \|y\|^2),$$

we find that

$$(Tx, y) + (Ty, x) \leq \frac{1}{2} M_T(\|x + y\|^2 + \|x - y\|^2)$$
$$= M_T(\|x\|^2 + \|y\|^2).$$

Now, if $Tx = 0$, it is clear that $\|Tx\| \leq M_T\|x\|$, so assume that $Tx \neq 0$ and define

$$y = \frac{\|x\|}{\|Tx\|} Tx.$$

Then $\|x\| = \|y\|$ and

$$2\frac{\|x\|}{\|Tx\|}\|Tx\|^2 = \frac{\|x\|}{\|Tx\|}((Tx, Tx) + (TTx, x))$$
$$= (Tx, y) + (Ty, x)$$
$$\leq M_T(\|x\|^2 + \|y\|^2)$$
$$= 2M_T\|x\|^2,$$

from which we conclude that $\|Tx\| \leq M_T\|x\|$ for all $x \in H$, hence $\|T\| \leq M_T$, and the result follows.   ∎

In finite dimensions we know that a linear operator in C^k is injective if and only if it is surjective, hence invertibility is equivalent to surjectivity. For self-adjoint operators we have a corresponding result.

PROPOSITION 5.5
Let T be a bounded, self-adjoint operator on a Hilbert space H. If $T(H)$ is dense in H, then T has an inverse .

PROOF   Since $\overline{T(H)} = H$, we have that $\{0\} = ker(T^*) = ker(T)$, so T is injective, and the inverse $T^{-1} : T(H) \to H$ is well defined.   ∎

REMARK 5.2   The inverse operator above is not necessarily bounded, in fact, in most applications (differential equations formulated as integral equations) it will be unbounded.   ∎

PROPOSITION 5.6
Assume that P is a bounded, self-adjoint operator on the Hilbert space H, and assume that $P^2 = P$.
   Then P is the orthogonal projection on the closed subspace

$$M = \{z \in H \mid Pz = z\}. \tag{5.7}$$

PROOF

Notice that $\|Px\|^2 = (Px, Px) = (P^2x, x) = (Px, x) \leq \|Px\|\|x\|$, so $\|P\| \leq 1$. Since P is bounded, the set $M = ker(P - I)$ is closed, and we see that $M = P(H)$ since if $y \in P(H)$, then there is an $x \in H$ with $y = Px$, so $Py = P^2x = Px = y$, implying that $y \in M$. On the other hand, if $y \in M$, then $y = Py$, and obviously $y \in P(H)$.

If $x \in M^\perp$, then $(Px, y) = (x, Py) = (x, y) = 0$ for all $y \in M$, hence $Px \in M^\perp$, but then $Px = 0$. So if $z = x + y$ where $x \in M$ and $y \in M^\perp$, then $Pz = x$, and P is the orthogonal projection on M. ∎

5.2 Compactness and Compact Operators

We will now introduce a special class of bounded linear operators that occur in many applications, in particular in integral equations or as inverses of unbounded operators. Historically, the so called *compact* operators were the first to be studied as they enjoy many properties in common with the linear operators on finite dimensional spaces studied in linear algebra. In a certain sense, the study of compact operators was the very start of modern functional analysis. The compact operators are sometimes called the *completely continuous* operators for reasons that will become clear.

Let us recall the definition of a compact set from Chapter 2: a set S in a normed space is *compact* if every sequence in S has a subsequence that converges to an element of S.

From elementary calculus we know that in *finite* dimensions, the compact sets are precisely the closed and bounded ones. This is not true in infinite dimensions; as an example, take the closed unit ball $\{x \in H \mid \|x\| \leq 1\}$ in a Hilbert space H. Then any orthonormal sequence (x_n) will belong to it, but since $\|x_n - x_m\|^2 = 2$ for all $n \neq m$, no subsequence can converge. On the other hand, we have:

PROPOSITION 5.7

A compact set is closed and bounded.

PROOF Let S be a set. If S is not bounded, we can to every $n \in N$ take an x_n in S, with $\|x_n\| > n$. For every subsequence (x_{n_k}) of (x_n) we have that $\|x_{n_k}\| > n_k$, hence $\|x_{n_k}\| \to \infty$, so (x_{n_k}) is not convergent and S is not compact.

If S is not closed, there is an $x \in \overline{S} \setminus S$ and a sequence (x_n) in S that converges to x. But then any subsequence of (x_n) will converge to $x \notin S$, hence S is not compact.   ∎

A linear operator is bounded if it maps bounded sets into bounded sets. A compact operator has even stronger continuity properties:

DEFINITION 5.2   *Let V and W be normed spaces. An operator $T \in B(V, W)$ is compact if $\overline{T(A)}$ is compact in W for all bounded sets $A \subset V$.*

Notice that in an infinite dimensional Hilbert space, the unit ball was not compact, hence the identity operator I is not a compact operator.

Example 5.3
If T is a bounded operator on a normed space V, we say that T has finite rank if $dim T(V)$ is finite. Since a closed and bounded set in finite dimensions is compact, we see that operators of finite rank are compact.
⬚

In a moment we shall see that the compact operators in $B(V, W)$ are exactly the closure of the set of operators of finite rank. Since the concept of weak convergence in a Hilbert space coincides with usual convergence whenever the space is finite dimensional, it is natural to guess that compact operators have special properties with respect to weak convergence.

PROPOSITION 5.8
Let H be a Hilbert space and let (x_n) be a weakly convergent sequence, with weak limit x. If $T \in B(H)$ is compact, then (Tx_n) converges in norm to Tx.

PROOF   Since $x_n \rightharpoonup x$ we have for all $y \in H$ that

$$(Tx_n, y) = (x_n, T^*y) \to (x, T^*y) = (Tx, y) \quad \text{for} \quad n \to \infty,$$

hence (Tx_n) converges weakly to Tx. Since strong convergence implies weak convergence, we see that Tx is the only possible limit. So assume that (Tx_n) does not converge to Tx. Then it is possible to extract a subsequence (Tx_{n_k}) of (Tx_n) such that

$$\|Tx_{n_k} - Tx\| > \delta \quad \text{for all} \quad k \in N$$

and some $\delta > 0$. But since (x_n) is bounded and T is compact, we can find a subsequence $(Tx_{n_{k_l}})$ of (Tx_{n_k}) that is strongly convergent to a y in H, but since $Tx_{n_{k_l}} \rightharpoonup Tx$ we must have that $y = Tx$, which is not possible according to the inequality above. Hence $Tx_n \to Tx$.   ∎

If (e_n) is an orthonormal sequence, the proposition above implies that $Te_n \to 0$, since $e_n \rightharpoonup 0$. This implies the following proposition.

PROPOSITION 5.9

Let H be an infinite dimensional Hilbert space. If $T \in B(H)$ is compact and T^{-1} exists, then T^{-1} is unbounded.

PROOF   Let (e_n) be an orthonormal sequence in H. Then $Te_n \to 0$ but $\|T^{-1}(Te_n)\| = \|e_n\| = 1$ for all n, so T^{-1} is not continuous.   ■

Now we will show that the set of compact operators is closed in $B(H)$ (in the operator norm).

THEOREM 5.10

Let H be a Hilbert space and assume that (T_n) is a sequence of compact operators in $B(H)$ converging to an operator T. Then T is compact.

PROOF   Let (y_n) be a bounded sequence in $T(H)$. Then there is a (possibly nonunique) bounded sequence (x_n) in H with $Tx_n = y_n$. We will show that it is possible to extract a subsequence (x_{n_k}) such that (Tx_{n_k}) is convergent.

Since T_1 is compact, (x_n) has a subsequence (x_n^1) such that $(T_1 x_n^1)$ is convergent. Since T_2 is compact, (x_n^1) has a subsequence (x_n^2) such that $(T_2 x_n^2)$ is convergent. Continuing this way gives us, for every k, a subsequence (x_n^k) of (x_n) such that $(T_k x_n^k)$ is convergent. Consider now the *diagonal sequence* (x_n^n). It is obvious that $(T_k x_n^n)$ is convergent for every k, and we will show that (Tx_n^n) is a Cauchy sequence in H.

Let $\epsilon > 0$ be given. Since (x_n) is bounded, there is a positive constant C such that $\|x_n^n\| \le C$ for every n. Since $T_n \to T$, we can find a fixed number k such that

$$\|T - T_k\| \le \frac{\epsilon}{3C}.$$

Since $(T_k x_n^n)$ is convergent, there is a number N such that

$$\|T_k x_n^n - T_k x_m^m\| \le \frac{\epsilon}{3}$$

for all $m, n > N$.

Now, for $m, n > N$ we have

$$\|Tx_n^n - Tx_m^m\| \leq \|Tx_n^n - T_k x_n^n\| + \|T_k x_n^n - T_k x_m^m\| + \|T_k x_m^m - Tx_m^m\|$$

$$\leq \|T - T_k\|\|x_n^n\| + \frac{\epsilon}{3} + \|T_k - T\|\|x_m^m\|$$

$$\leq \frac{\epsilon}{3C}C + \frac{\epsilon}{3} + \frac{\epsilon}{3C}C$$

$$= \epsilon.$$

Since (Tx_n^n) is a Cauchy sequence in H, it is convergent, and since the diagonal sequence is a subsequence of (x_n), the proof is complete. ∎

PROPOSITION 5.11

Let (e_n) be an orthonormal basis for a Hilbert space H and let (λ_n) be a sequence of complex numbers. Define the operator $T : H \to H$ by

$$Tx = \sum_{n=1}^{\infty} \lambda_n (x, e_n) e_n.$$

Then T is bounded if and only if (λ_n) is bounded; and T is compact if and only if $\lambda_n \to 0$.

PROOF

The first statement is trivial since $\|Tx\|^2 = \sum_{n=1}^{\infty} |\lambda_n|^2 |(x, e_n)|^2$ and $\|x\|^2 = \sum_{n=1}^{\infty} |(x, e_n)|^2$. To prove the second statement, assume that $\lambda_n \to 0$. We define the sequence of operators (T_k) by

$$T_k x = \sum_{n=1}^{k} \lambda_n (x, e_n) e_n. \tag{5.8}$$

Since T_k has finite rank for every k, all T_k's are compact, and since

$$\|Tx - T_k x\|^2 = \|\sum_{n=k+1}^{\infty} |\lambda_n|^2 |(x, e_n)|^2 \leq K_k \|x\|^2, \tag{5.9}$$

where $K_k = \sup_{n>k}\{|\lambda_n|^2\}$ we see that

$$\|T - T_k\| \leq \sqrt{K_k} \to 0, \tag{5.10}$$

implying that $T_k \to T$, so T is compact. On the other hand, assume that $\lambda_n \not\to 0$. Then there is a subsequence (λ_{n_k}) of (λ_n) and an $\epsilon > 0$ such that $|\lambda_{n_k}| \geq \epsilon$ for all k. Consider the corresponding subsequence (e_{n_k}) of (e_n). It is orthonormal, hence weakly convergent to 0, but since

$$\|Te_{n_i} - Te_{n_j}\|^2 = \|\lambda_{n_i} e_{n_i} - \lambda_{n_j} e_{n_j}\|^2 = |\lambda_{n_i}|^2 + |\lambda_{n_j}|^2 > 2\epsilon^2$$

for all $i \neq j$, (Te_n) can have no subsequences that are Cauchy, hence (Te_{n_k}) does not converge to 0, and T is not compact. ∎

5.3   Closed Operators

We have seen that the bounded, linear operators on a Hilbert space enjoy many nice properties but, unfortunately, many of the linear operators we meet in applications are not bounded. We have so far only considered linear operators defined on the entire Hilbert or Banach space, but when the operators are not bounded, the *domain of definition*, usually denoted $D(T)$ for an operator T, becomes extremely significant. As an example, consider the Hilbert space $L^2(I)$ where I is a bounded interval, and let $T = \frac{d}{dx}$. Then we can take $D(T) = C^1(I)$ or $D(T) = C^2(I)$ and notice that even if the *action* of the operator is the same on the two domains, it is *two different operators*; one is obviously a restriction of the other. So when we are dealing with unbounded operators, it is important to specify both the action and the domain. Recall that from the definition of a linear operator, the domain of a linear operator has to be a linear subspace. This calls for some definitions.

DEFINITION 5.3   *Let T_1 and T_2 be linear operators from a normed space V into a normed space W, with domains $D(T_1)$ and $D(T_2)$, respectively.*
   We say that $T_1 = T_2$ if

$$D(T_1) = D(T_2) \quad \text{and} \quad T_1 x = T_2 x \quad \text{for} \quad x \in D(T_1).$$

   We say that T_1 is a restriction *of T_2 or, equivalently, T_2 is an extension of T_1, if*

$$D(T_1) \subset D(T_2) \quad \text{and} \quad T_1 x = T_2 x \quad \text{for} \quad x \in D(T_1),$$

and we write $T_1 \subset T_2$ in this case.
   Moreover, we define the sum *of T_1 and T_2 , $T_1 + T_2$ by*

$$D(T_1 + T_2) = D(T_1) \cap D(T_2) \quad \text{and}$$
$$(T_1 + T_2)x = T_1 x + T_2 x \quad \text{for} \quad x \in D(T_1 + T_2).$$

Example 5.4
We must be careful with these definitions. So let $V = W = L^2(J)$ where J is a bounded interval, and let T_1 be the operator

$$D(T_1) = C^1(J) \quad \text{and} \quad T_1 f = \frac{d}{dx} f \quad \text{for} \quad f \in D(T_1).$$

Let I be the identity operator

$$D(I) = L^2(J) \quad \text{and} \quad If = f \quad \text{for} \quad f \in D(I),$$

and let us consider the operator $T_2 = I - T_1$. According to the definition
we have

$$D(T_2) = D(T_1) \cap D(I) = C^1(J) \quad \text{and} \quad T_2 f = f - \frac{d}{dx} f \quad \text{for} \quad f \in C^1(J).$$

If we consider the operator $T_1 + T_2$, we find that

$$D(T_1 + T_2) = D(T_1) \cap D(T_2) = C^1(J) \subset L^2(J)$$

and

$$(T_1 + T_2)f = T_1 f + T_2 f = \frac{d}{dx} f + (f - \frac{d}{dx} f) = f \quad \text{for} \quad f \in C^1(J),$$

hence $T_1 + T_2 \subset I$, and $T_1 + T_2$ is a proper restriction of the identity.
$\square$

Also, the *composition* $T_2 T_1$ of two operators calls for some comments; we
define

$$D(T_2 T_1) = \{x \in D(T_1) \mid T_1 x \in D(T_2)\}$$

and

$$T_2 T_1 x = T_2(T_1 x) \quad for \quad x \in D(T_2 T_1).$$

In this way we can define powers and polynomials of unbounded operators.

As a famous example, let us consider what is known as *the Heisenberg
Uncertainty Relation*:

Example 5.5
Let $H = L^2([0; 1])$ and consider the following two operators, ∂ and M, from
H into H,

$$D(\partial) = C^1([0; 1]) \quad \text{and} \quad \partial f = \frac{d}{dx} f \quad \text{for} \quad f \in D(\partial),$$

and

$$D(M) = H \quad \text{and} \quad M f(x) = x f(x) \quad \text{for} \quad f \in H.$$

Now, M is in fact bounded, since $\|Mf\|^2 = \int_0^1 |xf(x)|^2 dx \leq \|f\|^2$, so
it is obvious that $D(\partial M) = D(M\partial) = C^1([0; 1])$. We can then define the
commutator $[\partial, M]$ of ∂ and M by

$$[\partial, M] = \partial M - M\partial \quad \text{with} \quad D([\partial, M]) = C^1([0; 1]).$$

Notice that if $[A, B] \subset 0$ for two operators A and B, then they commute.
But

$$[\partial, M]f(x) = (\partial M - M\partial)f(x) = \partial(xf(x) - x(\partial f(x)) = f(x) \quad \text{for} \quad f \in D([\partial, M]),$$

and we see that $[\partial, M] \subset I$, which (apart from a constant) is the Heisenberg Uncertainty Relation from quantum mechanics. $\Box$

We have seen that the differential operators, like ∂ above, must be treated with great care. In order to deal with these operators (and other unbounded operators) we will now consider the class of operators that have a so-called *closed extension*, but first we need some definitions. If V and W are normed spaces, then the *product space* $V \times W$, with the obvious rules $(x_1, y_1) + (x_2, y_2) = (x_1 + x_2, y_1 + y_2)$, $\lambda(x, y) = (\lambda x, \lambda y)$, is also a normed space (be careful with the notation, (x,y) is an ordered pair, not an inner product). The norm in $V \times W$ is the *product norm*

$$\|(x, y)\| = \|x\| + \|y\|, \tag{5.11}$$

where the norm symbol $\| \cdot \|$ is used in three different ways. If V and W are Banach spaces, so is $V \times W$ (see Exercise 33).

Now, let V and W be normed spaces, and T a linear operator from V into W, with $D(T) \subset V$. We can then consider the *graph* of T, denoted $G(T)$ and defined by

$$G(T) = \{(x, Tx) \mid x \in D(T)\} \subset V \times W. \tag{5.12}$$

Notice that $G(T)$ is a linear subspace of $V \times W$ since T is linear.

DEFINITION 5.4   *Let $T : V \to W$ be a linear operator with domain $D(T)$. We will say that T is a* closed *operator if $G(T)$ is closed in $V \times W$ with respect to the product norm.*

We will now prove a number of famous theorems from operator theory that are very frequently used, ending with the so-called *closed graph theorem*. We start with a technical lemma where we will use the notation of *open balls* $B_X(x_0, r)$, defined by

$$B_X(x_0, r) = \{x \in X \mid \|x - x_0\| < r\}. \tag{5.13}$$

LEMMA 5.12
Let T be a bounded operator from a Banach space V into a Banach space W, with $D(T) = V$.
Assume that the closure of the set $T(B_V(0, 1))$ contains an open ball $B_W(0, r)$ for some $r > 0$.
Then $B_W(0, r) \subset T(B_V(0, 1))$.

PROOF   Our assumption implies that the set $T(B_V(0, 1))$ must be dense in $B_W(0, r)$, so any element in $B_W(0, r)$ can be approximated with a sequence from $T(B_V(0, 1))$.

Now, let $y \in B_W(0, r)$. We will find an $x \in B_V(0, 1)$ such that $Tx = y$.

Since $\|y\| < r$, we can find a δ, $0 < \delta < 1$ such that $\|y\| < (1 - \delta)r = r_1$, hence $y \in B_W(0, r_1)$. Since T is linear, $(1 - \delta)T(B_V(0, 1))$ is dense in $B_W(0, r_1) = (1 - \delta)B_W(0, r)$.

We define now $y_0 = 0 \in T(B_V(0, 1))$. It is trivial that the set $B_W(y_0, r_1) \cap B_W(y, \delta r_1)$ is nonempty, since y lies in both balls. Now, since $(1-\delta)T(B_V(0, 1))$ is dense in $B_W(0, r_1) = B_W(y_0, \delta^0 r_1)$, we have obviously a $y_1 \in B_W(y_0, \delta^0 r_1) \cap B_W(y, \delta r_1)$ such that $y_1 - y_0 \in \delta^0(1 - \delta)T(B_V(0, 1))$.

Now we proceed by induction, so assume that we have determined a sequence $y_0, y_1, ..., y_n$ such that

$$y_k \in B_W(y_{k-1}, \delta^{k-1} r_1) \cap B_W(y, \delta^k r_1) \tag{5.14}$$

and

$$y_k - y_{k-1} \in \delta^{k-1}(1 - \delta)T(B_V(0, 1)) \tag{5.15}$$

for $k = 1, 2, ..., n$.

We will find $y_{n+1} \in W$ satisfying the above with $k = n + 1$.

Since $y_n \in B_W(y, \delta^n r_1)$, we have that $\|y - y_n\| < \delta^n r_1$, so the set

$$B_W(y_n, \delta^n r_1) \cap B_W(y, \delta^{n+1} r_1) \tag{5.16}$$

cannot be empty since y is in this intersection of open balls. But since $\delta^n(1 - \delta)T(B_V(0, 1))$ is dense in the set $\delta^n(1 - \delta)B_W(0, r) = B_W(0, \delta^n r_1)$, we can find a $y_{n+1} \in B_W(y_n, \delta^n r_1) \cap B_W(y, \delta^{n+1} r_1)$, with $y_{n+1} - y_n \in \delta^n(1 - \delta)T(B_V(0, 1))$. Then, by induction, we have a sequence (y_n) satisfying the inclusions above for all $k \in N$.

Now, in particular, $y_n \in B_W(y, \delta^n r_1)$, and since $\delta^n r_1 \to 0$ for $n \to \infty$, we see that $y_n \to y$ in W for $n \to \infty$.

Moreover, since $y_{n+1} - y_n \in \delta^n(1-\delta)T(B_V(0, 1)) = T(B_V(0, \delta^n(1-\delta)))$, there is an $x_{n+1} \in B_V(0, \delta^n(1 - \delta))$ such that

$$Tx_{n+1} = y_{n+1} - y_n \tag{5.17}$$

for $n \in N$.

If we denote

$$s_N = \sum_{n=1}^{N} x_n = \sum_{n=0}^{N-1} x_{n+1} \tag{5.18}$$

and

$$x = \lim_{N \to \infty} s_N, \tag{5.19}$$

we have that

$$\|x\| \le \sum_{n=0}^{\infty} \|x_{n+1}\| < \sum_{n=0}^{\infty} \delta^n (1-\delta) = 1, \qquad (5.20)$$

so $x \in B_V(0,1)$ and, by the continuity of T, $Ts_N \to Tx$ for $N \to \infty$.
On the other hand,

$$Ts_N = \sum_{n=0}^{N-1} Tx_{n+1} = \sum_{n=0}^{N-1} (y_{n+1} - y_n) = y_N - y_0 = y_N,$$

where we saw that $y_N \to y$ for $N \to \infty$. So the sequence (Ts_N) converges to both Tx and y, hence $Tx = y$, and the result follows. ∎

This lemma is one of the versions of the so called *Open mapping theorem.* We will proceed immediately to the next famous theorem:

THEOREM 5.13 (Bounded inverse theorem)
Let $T : V \to W$ be a bounded, linear, and bijective operator from the Banach space V onto the Banach space W. Then the inverse operator $T^{-1} : W \to V$ is also bounded.

PROOF From the assumptions, $T(B_V(0,1))$ must be dense in some open ball $B \subset W$, so we can take a $y_0 \in B \cap T(B_V(0,1))$ such that there is a corresponding $x_0 \in B_V(0,1)$ with $Tx_0 = y_0$. If we define the open ball B_1 by

$$B_1 = B - y_0 = \{y - y_0 \mid y \in B\}, \qquad (5.21)$$

we have that $0 \in B_1$, so we can find an $r > 0$ such that $B_W(0,r) \subset B_1$. It is easy to see that

$$B_V(0,1) - x_0 \subset B_V(0,2) \qquad (5.22)$$

so, from the linearity of T we find that

$$T(B_V(0,1)) - y_0 = T(B_V(0,1)) - Tx_0 \subset T(B_V(0,2)). \qquad (5.23)$$

Since $T(B_V(0,1))$ is dense in B, $T(B_V(0,2))$ must be dense in $B - y_0 = B_1$, hence in $B_W(0,r)$ in particular. By the linearity, $T(B_V(0,1))$ must then be dense in $B_W(0,\frac{r}{2})$, and from the preceding lemma we must have that

$$B_W(0,\frac{r}{2}) \subset T(B_V(0,1)) \qquad (5.24)$$

or, by linearity again,

$$Bw(0,1) \subset T(B_V(0, \frac{2}{r})). \tag{5.25}$$

Since T is bijective, this implies that

$$T^{-1}(Bw(0,1)) \subset (B_V(0, \frac{2}{r})), \tag{5.26}$$

which is just the statement that $\|T^{-1}\| \leq \frac{2}{r}$, that is, T^{-1} is bounded.   ∎

THEOREM 5.14 (Closed graph theorem)
Let V and W be Banach spaces and let $T : D(T) \subset V \to W$ be a closed linear operator. If the domain $D(T)$ is closed in V, the operator T is bounded.

PROOF   From the assumptions, both $D(T)$ and the graph $G(T)$ are Banach spaces. We define an operator $P : G(T) \to D(T)$ by

$$P(x, Tx) = x. \tag{5.27}$$

Then P is linear, and since

$$\|P(x, Tx)\| = \|x\| \leq \|x\| + \|Tx\| = \|(x, Tx)\|,$$

the operator P is bounded. The inverse operator P^{-1} is given by

$$P^{-1}x = (x, Tx) \quad \text{for} \quad x \in D(T). \tag{5.28}$$

Now, from the bounded inverse theorem, P^{-1} is a bounded operator from $D(T)$ onto $G(T)$, so

$$\|P^{-1}x\| = \|(x, Tx)\| \leq \|P^{-1}\|\|x\|, \tag{5.29}$$

which implies that T must be bounded, since

$$\|Tx\| \leq \|x\| + \|Tx\| = \|(x, Tx)\|. \tag{5.30}$$

∎

The theorem above applies to closed operators, so it is important to know when an operator is closed or, if not, has an extension that is.

So, let V and W be Banach spaces, and $T : V \to W$ be a linear operator with domain $D(T) \subset V$. We will say that T is *closeable*, or that T has a *closed extension*, if the closure of the graph $G(T)$ in $V \times W$, $\overline{G(T)}$ is the graph of an operator $\overline{T}$. (That is, if $\overline{G(T)} = G(\overline{T})$.) If $\overline{T}$ exists, it is called the *closure* of T. Since $G(T) \subset G(\overline{T})$, it is clear that $T \subset \overline{T}$.

PROPOSITION 5.15

A linear operator T is closable if and only if $\overline{G(T)}$ does not contain a pair of the form $(0, y) \in V \times W$ where $y \neq 0$.

PROOF If T is closeable, $\overline{G(T)}$ is the graph of the closure $\overline{T}$. Since $0 \in D(T) \subset D(\overline{T})$, we see that $\overline{T}0 = 0$, so for all $y \neq 0$, the pair $(0, y)$ can never belong to $G(\overline{T}) = \overline{G(T)}$.

If T is not closeable, then $\overline{G(T)}$ is not a graph. This means that $\overline{G(T)}$ contains pairs of the form (x, y_1) and (x, y_2) where $y_1 \neq y_2$. Since $\overline{G(T)}$ is a linear subspace of $V \times W$, we see that the pair $(0, y_1 - y_2) \in \overline{G(T)}$. Since $y_1 - y_2 \neq 0$, the proposition follows. ∎

We have the following characterization of closed operators:

PROPOSITION 5.16

The linear operator $T : V \to W$ is closed if and only if:
(x_n) is a sequence in $D(T)$ such that

$$x_n \to x \in V \quad \text{and} \quad Tx_n \to y \in W,$$

then

$$x \in D(T) \quad \text{and} \quad y = Tx.$$

PROOF A pair (x, y) is in $\overline{G(T)}$ if and only if there is a sequence (x_n, Tx_n) in $G(T)$, such that

$$(x_n, Tx_n) \to (x, y) \quad \text{for} \quad n \to \infty,$$

that is, if $x_n \to x$ and $Tx_n \to y$ for $n \to \infty$. We see that $\overline{G(T)} = G(T)$ if and only if $(x, y) \in G(T)$, and in this case $(x, y) = (x, Tx)$, so $x \in D(T)$ and $y = Tx$. ∎

Notice the important difference to what we know about bounded operators. A bounded operator T can *always* be closed if it is not so from the start, because we can, without any problems, extend T to $\overline{D(T)}$; by the continuity of T, the sequence (Tx_n) will always converge if (x_n) does. So the concept of closure is only relevant when dealing with unbounded, linear operators. The most interesting unbounded operators are the differential operators, and these share the property that almost any natural domain we can think of will be dense in the Hilbert space L^2 (over an interval or R if the dimension is 1). Such operators are called *densely defined*.

Example 5.6
Let us consider the differential operator $\partial = \frac{d}{dx}$ in $L^2(I)$, where we take $I = [0; 1]$ for convenience. Let $D(\partial) = C^1(I) \subset L^2(I)$. Then ∂ is obviously linear and $\partial : D(\partial) \to C^0(I) \subset L^2(I)$. Let us convince ourselves that ∂ is unbounded. We define the sequence (f_n) in $D(\partial)$ by

$$f_n(x) = \frac{1}{n} \sin(2n\pi x), \tag{5.31}$$

so

$$\partial f_n(x) = 2\pi \cos(2n\pi x) \tag{5.32}$$

and

$$\begin{aligned}
\|\partial f_n\|_2^2 &= 4\pi^2 \int_0^1 \cos^2(2n\pi x)dx \\
&= 4\pi^2 \int_0^1 \sin^2(2n\pi x)dx \\
&= \frac{1}{2} \cdot 4\pi^2 (\int_0^1 (\sin^2(2n\pi x) + \cos^2(2n\pi x))dx) \\
&= 2\pi^2.
\end{aligned}$$

But since

$$\|f_n\|_2^2 = \frac{1}{n^2} \int_0^1 \sin^2(2n\pi x)dx = \frac{1}{2n^2}, \tag{5.33}$$

we see that $f_n \to 0$ in $L^2(I)$ and $\partial f_n \nrightarrow 0$, so ∂ must be unbounded.

Let us show that ∂ with $D(\partial) = C^1(I)$ is closeable. So assume that a sequence (g_n) in $D(\partial)$ is chosen such that $g_n \to 0$ in $L^2(I)$ and $\partial g_n \to g$ in $L^2(I)$. We must then show that $g = 0$.

Since the inner product is continuous, we have that

$$(\partial g_n, \varphi) \to (g, \varphi) \quad \text{for all} \quad \varphi \in C_0^\infty(I), \tag{5.34}$$

and we have also that

$$(\partial g_n, \varphi) = \int_0^1 \partial g_n(x)\overline{\varphi(x)}dx \tag{5.35}$$

$$= 0 - \int_0^1 g_n(x)\overline{\partial\varphi(x)}dx \tag{5.36}$$

$$= -(g_n, \partial\varphi(x)) \tag{5.37}$$

$$\to -(0, \partial\varphi(x)) \tag{5.38}$$

$$= 0, \tag{5.39}$$

from which we conclude that $(g, \varphi) = 0$ for all $\varphi \in C_0^\infty(I)$. This means that $g \in C_0^\infty(I)^\perp$, but since $C_0^\infty(I)$ is dense in $L^2(I)$, we have that $C_0^\infty(I)^\perp = L^2(I)^\perp = \{0\}$, so $g = 0$. This shows that ∂ with $D(\partial) = C^1(I)$ is closeable, with a closure $\overline{\partial}$ that has a domain $D(\overline{\partial}) \subset L^2(I)$ consisting of functions that can be differentiated in some sense. More precisely, $D(\overline{\partial})$ is the functions $f \in L^2(I)$ for which there exist a sequence (f_n) in $C^1(I)$ with $f_n \to f$ and ∂f_n convergent in $L^2(I)$, and $\overline{\partial} f = \lim_n \partial f_n$. We will find out later on that $D(\overline{\partial})$ is the so-called *Sobolev Space* $H^1(I)$, which is very important in applications. $\quad \square$

5.4 The Adjoint of an Unbounded Operator

In the proof of the existence of an adjoint for a bounded operator, the Riesz' representation theorem was the essential ingredient, since the functional $\varphi_y(x) = (Tx, y)$ was *bounded* for all x and y in H. When T is not bounded, we can of course define the functional $\varphi_y(x) = (Tx, y)$ for $x \in D(T) \subset H$, but it is not bounded on H and it is not clear *if* we can find $z \in H$ such that $(Tx, y) = (x, z)$ for all $x \in D(T)$. Now assume that $D(T)$ is dense in H and let $x \in D(T)$. If $y \in H$ has the property that there exists a $z \in H$ such that $(Tx, y) = (x, z)$, we see that this $z \in H$ is unique by the density of $D(T)$. (Since $(x, z_1) = (x, z_2)$ implies that $z_1 - z_2 \in D(T)^\perp = \{0\}$.) But then the pair (y, z) (not inner product) belongs to the graph of an operator which we will denote T^*. We have then

DEFINITION 5.5 *Let T be a densely defined linear operator in the Hilbert space H. The* adjoint *of T is the operator T^*, given by*

$$(Tx, y) = (x, T^*y) \quad \text{for} \quad x \in D(T) \quad \text{and} \quad y \in D(T^*)$$

where $D(T^*)$ is the subspace of H for which there exists a $z \in H$ satisfying $(Tx, y) = (x, z)$ whenever $y \in D(T^*)$ and $x \in D(T)$, and then $T^*y = z$.
If $T = T^*$ we say that T is self-adjoint.

REMARK 5.3 If T is densely defined and T_1 is an extension of T, then T^* is an extension of T_1^*, since for $y \in D(T_1^*)$ and $x \in D(T)$ we have

$$(Tx, y) = (T_1x, y) = (x, T_1^*y),$$

so $y \in D(T^*)$ with $T^*y = T_1^*y$. $\quad \blacksquare$

PROPOSITION 5.17

If T is a densely defined operator, T^* is a closed linear operator.

PROOF   It is trivial that $D(T^*)$ is a subspace of H and $T^*(\alpha x + \beta y) = \alpha T^*(x) + \beta T^*(y)$ for $x, y \in D(T^*)$.

If (x_n) is a sequence in $D(T^*)$ converging to x such that (T^*x_n) converges to y, then for $z \in D(T)$ we have $(z, T^*x_n) = (Tz, x_n) \to (Tz, x)$, and we see that $(Tz, x) = (z, y)$ for each $z \in D(T)$. Hence $x \in D(T^*)$ with $T^*x = y$, and T^* is closed.   ∎

If T was a bounded operator on the Hilbert space H, we saw earlier that

$$ker(T^*) = T(H)^\perp, \tag{5.40}$$

and we will now show that this also holds when T is unbounded and densely defined.

PROPOSITION 5.18

Assume that T is a densely defined operator on a Hilbert space H. Then

$$H = ker(T^*) \oplus \overline{T(D(T))}.$$

PROOF   If $y \in T(D(T))^\perp$, then for all $x \in D(T)$:

$$(Tx, y) = 0 = (x, 0),$$

so $y \in D(T^*)$ with $T^*y = 0$, hence $y \in ker(T^*)$.

On the other hand, if $y \in ker(T^*)$ so $T^*y = 0$, then

$$(x, T^*y) = 0$$

for all $x \in H$, hence for $x \in D(T)$ in particular. But then

$$0 = (x, T^*y) = (Tx, y)$$

for all $x \in D(T)$, so $y \in T(D(T))^\perp$. Hence

$$ker(T^*) = T(D(T))^\perp$$

and the result follows from the projection theorem since $ker(T^*)^\perp = \overline{T(D(T))}$.   ∎

Notice how this gives a nice description of the range $T(D(T))$ for a densely defined operator, which can be used in the discussion of the solution to the equation $Tx = f$ for $f \in H$. In order to solve this equation, it is obvious that f must belong to $\overline{T(D(T))}$, hence $(f, y) = 0$ for all y, satisfying $T^*y = 0$. So if T^* is *injective*, the equation $Tx = f$ is solvable for all $f \in H$. (Notice that T^* being injective does not necessarily imply that T is injective.) In the special case where T is self-adjoint and we have $ker(T)^{\perp} = \overline{T(D(T))}$ according to the theorem, this result is one of the versions of *Fredholm's alternative*: the equation $Tx = f$, where T is self-adjoint, is solvable for all $f \in H$ satisfying $(f, y) = 0$ for all $y \in D(T)$ with $Ty = 0$.

Example 5.7

Let T be the closed extension of the operator ∂ from before, with the domain $D(T) = H^1([0; 1])$ consisting of functions in $L^2([0; 1])$ that are differentiable in the "weak" sense as described in the example. Then T is densely defined (obviously $C_0^{\infty}([0; 1]) \subset D(T)$), so T^* exists. Arguing formally, we see that

$$(Tu, v) = \int_0^1 u'(x)\overline{v(x)}dx$$

$$= u(1)\overline{v(1)} - u(0)\overline{v(0)} - \int_0^1 u(x)\overline{v'(x)}dx,$$

hence $T^*v = -Tv$ with $D(T^*) = \{v \in H^1([0; 1]) \mid v(0) = v(1) = 0\}$.

Now consider the operator $T_1 \subset T$, where $D(T_1) = \{u \in H^1([0; 1]) \mid u(0) = 0\}$. If we then repeat the calculation above we see that $T_1^*v = -T_1v$ with $D(T_1^*) = \{v \in H^1([0; 1]) \mid v(1) = 0\}$.

Finally, consider the operator $T_2 \subset T$ where $D(T_2) = \{u \in H^1([0; 1]) \mid u(0) = u(1) = 0\} = D(T^*)$. In exactly the same way as before, we see that $T_2^*v = -T_2v$ and now $D(T_2^*) = D(T_2)$. This means that the operator iT_2 is self-adjoint since we have $(iT_2)^* = -iT_2^* = iT_2$.

Chapter 6

Spectral Theory

In this chapter we will use the theory developed so far to introduce the concept of eigenvalues and eigenvectors for linear operators on Hilbert spaces. From linear algebra we know how a self-adjoint matrix can be diagonalized; this means that there is an orthonormal basis $\{e_j\}$ for the (finite dimensional) vector space such that the linear mapping $f : C^k \to C^k$ corresponding to the matrix can be expressed as

$$f(x) = \sum_{j=1}^{k} \lambda_j(x, e_j)e_j.$$

Here we recognize $(x, e_j) = x_j, j = 1, 2, ..., k$ as the coordinates of x, and the $\lambda_j, j = 1, 2, ..., k$ are the *eigenvalues* with corresponding *eigenvectors* $e_j, j = 1, 2, ..., k$. Recall that these are the nontrivial solutions to the equation

$$f(x) = \lambda x.$$

We will examine the corresponding results for linear operators on infinite dimensional spaces, and as we have seen before, the infinite dimension causes not only technical problems. The finite dimensional results can be extended to the self-adjoint and compact operators for which we have the celebrated *spectral theorem*. If a vector x is the eigenvector for a linear mapping T that is injective with corresponding eigenvalue λ, then x is also eigenvector for T^{-1} with corresponding eigenvalue λ^{-1}. This fact gives us also a spectral theorem for the unbounded operators with compact and self-adjoint inverses, a property shared by a large class of interesting differential operators.

6.1   The Spectrum and the Resolvent

Let H be a Hilbert space and $T : H \to H$ a linear operator with $D(T) \subset H$. For $\lambda \in C$ we define the operator

$$T_\lambda = T - \lambda I \quad \text{and} \quad D(T_\lambda) = D(T);$$

here, I is the identity on H. Now let $y \in H$ and consider the equation

$$T_\lambda x = y.$$

If T_λ is *injective*, we can solve the equation to get

$$x = T_\lambda^{-1} y$$

for all $y \in T_\lambda(H) = D(T_\lambda^{-1})$. If T_λ^{-1} is also bounded, we see that it is only necessary for $T_\lambda(H) = D(T_\lambda^{-1})$ to be *dense* in H in order to solve the equation uniquely for *all* $y \in H$, since if this is the case (and $T_\lambda(H) \neq H$), any $y \in H \setminus T_\lambda(H)$ can be approximated by a sequence $(y_n) \subset T_\lambda(H)$. But $(T_\lambda^{-1} y_n)$ will converge, so we define $T_\lambda^{-1} y = \lim_n T_\lambda^{-1} y_n$ to be the solution.

The operator $T_\lambda^{-1} = (T - \lambda I)^{-1}$ is called the *resolvent* of T and is usually denoted $R_\lambda(T)$. It is defined for those $\lambda \in C$ where T_λ is injective, and if $R_\lambda(T)$ is also bounded and densely defined, it will have all the nice properties as a solution operator described above.

DEFINITION 6.1   *The* resolvent set *for T, $\rho(T)$, is the set of $\lambda \in C$ where $R_\lambda(T)$ exists as a densely defined and bounded operator on H. The complement $\sigma(T) = C \setminus \rho(T)$ is called the* spectrum *for T.*

If $\lambda \in \sigma(T)$, several things can be wrong with $R_\lambda(T)$. The most obvious case is when $(T - \lambda I)$ is not injective so $R_\lambda(T)$ does not even exist. This means that the equation

$$(T - \lambda I)x = 0,$$

has a nontrivial solution (or several), so

$$Tx = \lambda x,$$

and we call λ an *eigenvalue* of T, with corresponding *eigenvectors* x. Notice that the eigenvectors form a vector space (finite or infinite dimensional), called the *eigenspace* corresponding to λ.

DEFINITION 6.2   *The subset of $\sigma(T)$ consisting of eigenvalues of T is denoted the* point spectrum *of T.*

The continuous spectrum *for* T *is the subset of* $\sigma(T)$ *where* $R_\lambda(T)$ *exists as a densely defined but unbounded operator on* H.

The residual spectrum *for* T *is the subset of* $\sigma(T)$ *where* $R_\lambda(T)$ *exists, but is not densely defined.*

Example 6.1

If we consider a linear mapping from C^k into C^k with corresponding matrix T, we know that the roots $\lambda_j, j = 1, 2, ..., k$ (repeated according to multiplicity) of the characteristic polynomium for T are all eigenvalues for T, and since $R_\lambda(T)$ is in $B(C^k)$ when λ is not an eigenvalue, we see that the spectrum $\sigma(T)$ is a pure point spectrum, $\sigma(T) = \{\lambda_j\}_{j=1}^k$, and $\rho(T) = C \setminus \{\lambda_j\}_{j=1}^k$.   ▯

Example 6.2

Let $\partial = \frac{d}{dx}$, $D(\partial) = C^1([0; 1]) \subset H = L^2([0; 1])$. Then $(\partial - \lambda I)e^{\lambda x} = 0$ for all $\lambda \in C$, so $\sigma(\partial)$ is a pure point spectrum, $\sigma(T) = C$, thus $\rho(T) = \emptyset$. ▯

In the example above we saw that the unbounded operator ∂ had an empty resolvent set. This can never be the case for a bounded operator, as the next theorems will show.

PROPOSITION 6.1

Let H *be a Hilbert space and* $T \in B(H)$.

Then the resolvent set of T, $\rho(T)$, *is open and the spectrum,* $\sigma(T)$, *is closed.*

PROOF   If $\rho(T) = \emptyset$, the proposition is trivially true. (We will see in the next proposition that this can *never* be the case for a bounded operator.)

So assume $\rho(T) \neq \emptyset$ and take $\mu \in \rho(T)$. Then $R_\mu(T) = (T - \mu I)^{-1}$ is a densely defined, bounded operator, and for $\lambda \in C$ we have

$$T - \lambda I = T - \mu I - (\lambda - \mu)I = (T - \mu I)(I - (\lambda - \mu)R_\mu(T)).$$

Then, for

$$|\lambda - \mu| < \|R_\mu(T)\|^{-1},$$

we see that

$$\|(\lambda - \mu)R_\mu(T)\| < 1,$$

and we see that the operator $I - (\lambda - \mu)R_\mu(T)$ and hence $T - \lambda I$) also has a bounded inverse for $\lambda \in C$ with $|\lambda - \mu| < \|R_\mu(T)\|^{-1}$, which is an open ball with center μ. Thus the resolvent set contains an open ball around any of its points, so it is open and the complement $\sigma(T)$ is closed. ∎

PROPOSITION 6.2
Let H be a Hilbert space and $T \in B(H)$.

The spectrum $\sigma(T)$ of T is a compact set in C, contained in a circle with radius $\|T\|$ and center 0.

PROOF   Assume that $\lambda > \|T\|$.

Then $T - \lambda I = -\lambda(I - \frac{1}{\lambda}T)$ has a bounded inverse since $\|\frac{1}{\lambda}T\| < 1$, so $\lambda \in \rho(T)$.

This means that $\lambda \in \sigma(T)$ implies that $|\lambda| \leq \|T\|$, so the spectrum is bounded; since it is also closed and a subset of C, it is compact. ∎

Notice that $\lambda > \|T\|$ implies that $\lambda \in \rho(T)$ so that $\rho(T) \neq \emptyset$ when T is bounded.

For a bounded, self-adjoint operator T, we had that

$$\|T\| = \sup\{|(Tx, x)| \mid x \in H, \quad \|x\| = 1\}, \tag{6.1}$$

so we have the following

COROLLARY 6.3
Let T be a bounded, self-adjoint operator on a Hilbert space H. Then all $\lambda \in \sigma(T)$ satisfy

$$|\lambda| \leq \sup\{|(Tx, x)| \mid x \in H, \quad \|x\| = 1\}. \tag{6.2}$$

Example 6.3
Let us calculate the spectrum for the *one-sided shift operator* T: If (e_n) is an orthonormal basis for the Hilbert space H and

$$x = \sum_{n=1}^{\infty} \alpha_n e_n,$$

then

$$Tx = \sum_{n=2}^{\infty} \alpha_n e_{n-1}.$$

If $x = (x_1, x_2, ...)$ is an eigenvector with corresponding eigenvalue λ, then $Tx = \lambda x$, so

$$x_2 = \lambda x_1$$
$$x_3 = \lambda x_2 = \lambda^2 x_1 \quad \text{and so on,}$$
$$x_j = \lambda^{j-1} x_1.$$

Taking $x_1 = 1$ we have an eigenvector $x = (1, \lambda, \lambda^2, ...)$ provided that $x \in H$, that is if

$$\sum_{j=1}^{\infty} |\lambda^{j-1}|^2 < \infty,$$

showing that all $\lambda \in C : |\lambda| < 1$ is an eigenvalue. This is the *open* unit disk in C. Now, for all $x \in H$ we have

$$\|Tx\|^2 = \sum_{j=2}^{\infty} |x_{j-1}|^2 = \sum_{j=1}^{\infty} |x_j|^2 = \|x\|^2,$$

showing that $\|T\| = 1$. Then $\sigma(T)$ must be contained in the *closed* unit disk, and since $\sigma(T)$ is closed, it must be exactly the closed unit disk in C.
🗌

The self-adjoint operators are the generalizations of the self-adjoint matrices, and the next theorem is not surprising. Notice that the operator is not necessarily bounded.

THEOREM 6.4
If T is a self-adjoint operator on a Hilbert space, then all its eigenvalues are real, and eigenvectors corresponding to different eigenvalues are orthogonal.

PROOF   If $Tx = \lambda x$, then

$$\lambda\|x\|^2 = \lambda(x, x) = (Tx, x) = (x, Tx) = \overline{\lambda}(x, x) = \overline{\lambda}\|x\|^2,$$

and since $x \neq 0$, we see that $\lambda = \overline{\lambda}$, so λ is real. If also $Ty = \mu y$ we have that

$$\lambda(x, y) = (Tx, y) = (x, Ty) = \overline{\mu}(x, y) = \mu(x, y),$$

and since $\lambda \neq \mu$ we must have that $(x, y) = 0$.   ∎

THEOREM 6.5
Assume that T is a bounded and self-adjoint operator on a Hilbert space H. Then there is a sequence (x_n) in H with $\|x_n\| = 1$ for all n such that

$$(T - \lambda I)x_n \to 0 \quad \text{for} \quad n \to \infty,$$

where λ equals either $\|T\|$ or $-\|T\|$.

PROOF   Since $\|T\| = \sup\{|(Tx, x)| \mid x \in H, \quad \|x\| = 1\}$, there is a sequence (y_n) with $\|y_n\| = 1$ such that $|(Ty_n, y_n)| \to \|T\|$. Since (Ty_n, y_n)

is real, there is a subsequence (x_n) of (y_n) such that either $(Tx_n, x_n) \to \|T\|$ or $(Tx_n, x_n) \to -\|T\|$. Let $\lambda = \pm\|T\|$ be chosen such that $(Tx_n, x_n) \to \lambda$ for $n \to \infty$. Then we have

$$\begin{aligned} \|Tx_n - \lambda x_n\|^2 &= \|Tx_n\|^2 + \lambda^2\|x_n\|^2 - 2\lambda(Tx_n, x_n) \\ &\leq \|T\|^2 + \lambda^2 - 2\lambda(Tx_n, x_n) \\ &= 2\lambda^2 - 2\lambda(Tx_n, x_n) \to 2\lambda^2 - 2\lambda^2 = 0, \end{aligned}$$

for $n \to \infty$, hence $\|Tx_n - \lambda x_n\| \to 0$.

The theorem states that a self-adjoint operator has what could be called an *approximate* eigenvector; the only problem is that the sequence (x_n) does not necessarily converge. But when the operator is also compact, the sequence must converge. ∎

THEOREM 6.6

Assume that T is a compact and self-adjoint operator on a Hilbert space H. Then one of the numbers $\|T\|$ and $-\|T\|$ is an eigenvalue for T.

PROOF   The case where $T = 0$ is trivial, so assume that $T \neq 0$.

Since T is compact, the sequence (x_n) from the preceeding theorem has a subsequence (x_{n_k}) such that (Tx_{n_k}) converges. But $Tx_{n_k} - \lambda x_{n_k} \to 0$, so (x_{n_k}) must also converge to some $x_0 \in H$. Since $\|x_{n_k}\| = 1$, we see that $\|x_0\| = 1$ also. But then we have a nontrivial solution to the equation $Tx = \lambda x$, and λ is an eigenvalue. ∎

COROLLARY 6.7

Assume that T is a compact and self-adjoint operator on a Hilbert space H. Then the number

$$\max\{|(Tx, x)| \mid x \in H, \quad \|x\| = 1\} \tag{6.3}$$

exists and equals $\|T\|$. Moreover, the maximum is attained for an eigenvector (normalized) with corresponding eigenvalue either $\|T\|$ or $-\|T\|$.

PROPOSITION 6.8

Let T be a compact operator on a Hilbert space H, and assume that T has a nonzero eigenvalue λ. Then the corresponding eigenspace

$$H_\lambda = \{x \in H \mid Tx = \lambda x\}$$

is finite dimensional.

PROOF   Assume that H_λ is infinite dimensional. Then there is an orthonormal sequence $(x_n) \subset H_\lambda$ so that

$$\|Tx_n - Tx_m\|^2 = |\lambda|^2 \|x_n - x_m\|^2 = 2|\lambda|^2 > 0, \qquad (6.4)$$

for $n \neq m$, so (Tx_n) has no convergent subsequences and T cannot be compact. ∎

THEOREM 6.9 (Spectral Theorem for Compact Self-Adjoint Operators)

Let T be a compact and self-adjoint operator on a Hilbert space H. Then H has an orthonormal basis (e_n) consisting of eigenvectors for T. If H is infinite dimensional, the corresponding eigenvalues (different from 0) (λ_n) can be arranged in a decreasing sequence $|\lambda_1| \geq |\lambda_2| \geq \dots$ where $\lambda_n \to 0$ for $n \to \infty$, and if

$$x = \sum_n (x, e_n) e_n,$$

then

$$Tx = \sum_n \lambda_n (x, e_n) e_n.$$

PROOF   From the assumptions, T has an eigenvalue

$$\lambda_1 = \pm \max\{|(Tx, x)| \mid x \in H, \quad \|x\| = 1\}$$

and a corresponding normalized eigenvector, e_1. Let

$$Q_1 = \{e_1\}^\perp.$$

If $x \in Q_1$, then

$$(Tx, e_1) = (x, Te_1) = \lambda_1(x, e_1) = 0,$$

so $Tx \in Q_1$. Since Q_1 is an orthogonal complement, it is a closed subspace of H, hence Q_1 is a Hilbert space and we can consider T as a compact and self-adjoint operator on this Hilbert space. But then there is an eigenvalue $\lambda_2 = \pm \max\{|(Tx, x)| \mid x \in Q_1, \quad \|x\| = 1\}$, and it is clear that $|\lambda_1| \geq |\lambda_2|$. We call the corresponding normalized eigenvector for e_2, and it is clear that $e_1 \perp e_2$. Proceeding in this manner we get an orthonormal sequence of eigenvectors (e_n) and a decreasing sequence $|\lambda_1| \geq |\lambda_2| \geq \dots \geq |\lambda_n| \geq \dots$ and a sequence of subspaces $\dots \subset Q_n \subset Q_{n-1} \subset \dots \subset Q_1 \subset H$, each eigenvalue satisfying

$$|\lambda_n| = \max\{|(Tx, x)| \mid x \in Q_{n-1}, \quad \|x\| = 1\}.$$

If H is of finite dimension k, say, this procedure terminates after k steps, and we have an orthonormal basis consisting of k eigenvectors. If H is infinite dimensional, the sequence (e_n) converges weakly to 0, hence (Te_n) converges to 0 in norm, hence $|\lambda_n| = \|Te_n\| \to 0$ for $n \to \infty$.

Let M be the subspace of H consisting of all infinite linear combinations of the eigenvectors e_n. Then any $x \in H$ can be written in the form $x = z+y$, where $z \in M$ and $y \in M^\perp$. Since every element of $M^\perp$ is orthogonal to all the e_n, $M^\perp$ is contained in all the subspaces Q_n. We will show that $Ty = 0$ when $y \in M^\perp$. This is, of course, the case if $M^\perp = \{0\}$, else we write $y \in M^\perp$ as $y = \|y\|y_1$, where $y_1 = \frac{y}{\|y\|}$. Then

$$(Ty, y) = \|y\|^2 (Ty_1, y_1),$$

but since $y_1 \in Q_n$ for all n, we have that $|(Ty_1, y_1)| \leq |\lambda_n|$ for all n, so

$$|(Ty, y)| \leq \|y\|^2 |\lambda_n| \to 0,$$

since $|\lambda_n| \to 0$ for $n \to \infty$. Hence $(Ty, y) = 0$ implying that $Ty = 0$ on $M^\perp$. Then regarded as an operator on the Hilbert space $M^\perp$, T is the zero operator. Since $H = M \oplus M^\perp$, we can choose an orthonormal basis for $M^\perp$; each of these basis vectors will then be an eigenvector for T corresponding to the eigenvalue 0, and if we include these new eigenvectors in the sequence of eigenvectors constructed above, we have an orthonormal basis (e_n) for H consisting of eigenvectors for T. Since the new eigenvalue is 0, we still have that $\lambda_n \to 0$ for $n \to \infty$. Then, for any $x \in H$ we have that

$$x = \sum_{n=1}^{\infty} (x, e_n)e_n,$$

and by the continuity of T we have

$$Tx = T(\lim_n \sum_{k=1}^{n} (x, e_k)e_k) = \lim_n T(\sum_{k=1}^{n} (x, e_k)e_k)$$

$$= \lim_n \sum_{k=1}^{n} (x, e_k)Te_k$$

$$= \sum_{n=1}^{\infty} \lambda_n (x, e_n)e_n.$$

∎

The following theorem is now a corollary to the spectral theorem and the previous discussions.

THEOREM 6.10 (The Fredholm Alternative)
Let T be a compact and self-adjoint operator on the Hilbert space H, and consider the equation
$$(T - \lambda I)x = y$$
in H.

If λ is not an eigenvalue of T, then the equation has a unique solution
$$x = (T - \lambda I)^{-1}y$$

for all $y \in H$. If λ is an eigenvalue of T, then there is a solution if and only if $y \perp \ker(T - \lambda I)$, and in this case all solutions to the equation can be written in the form
$$x = x_0 + \sum_{i=1}^{n} \alpha_i e_i,$$

where x_0 is any solution to the equation, $\{e_1, e_2, ..., e_n\}$ is an orthonormal basis for the eigenspace $\ker(T - \lambda I)$, and the α_i are arbitrary complex numbers.

The following example illustrates how the Spectral Theorem and the Fredholm alternative together can be used to solve the equation $(T - \lambda I)x = y$, which we will later recognize as an abstract form of a *Fredholm Integral Equation*.

Example 6.4
Let T be a compact and self-adjoint operator on the Hilbert space H, and consider the equation
$$(T - \lambda I)x = y$$
in H. If λ is not an eigenvalue, we use the Spectral Theorem and write the equation
$$(T - \lambda I)x = \sum_{i=1}^{\infty}(\lambda_i - \lambda)(x, e_i)e_i = \sum_{i=1}^{\infty}(y, e_i)e_i,$$

where (λ_i) is the sequence of eigenvalues of T and (e_i) the corresponding eigenvectors.
We se that
$$(x, e_i) = (\lambda_i - \lambda)^{-1}(y, e_i),$$

so that the solution is
$$x = \sum_{i=1}^{\infty}(\lambda_i - \lambda)^{-1}(y, e_i)e_i.$$

☐

6.2   Operator-Valued Functions

As a first application of the spectral theorem we will give meaning to the operator $f(T)$, when f is a not too badly behaved function and T is compact and self-adjoint. Since the theory for self-adjoint matrices is trivial in this context, we will assume that the Hilbert spaces are infinite dimensional (but still separable).

PROPOSITION 6.11

Let T be a compact and self-adjoint operator on a Hilbert space H. Let (e_n) be an orthonormal basis for H consisting of eigenvectors for T, with corresponding (real) eigenvalues (λ_n).

Assume that $f : \sigma(T) \to C$ is a function such that the limit $\lim_n f(\lambda_n)$ for $n \to \infty$ exists.

Then the operator

$$f(T)x = \sum_{n=1}^{\infty} f(\lambda_n)(x, e_n)e_n \qquad (6.5)$$

is well defined, and $f(T)$ is compact if $\lim_n f(\lambda_n) = 0$, and self-adjoint if f is real valued.

PROOF   Write $\varphi = f - c$, where $\lim_n f(\lambda_n) = c$. Then the operator

$$\varphi(T)x = \sum_{n=1}^{\infty} \varphi(\lambda_n)(x, e_n)e_n$$

is well defined and compact since $\varphi(\lambda_n) \to 0$ for $n \to \infty$. Since

$$\sum_{n=1}^{k} f(\lambda_n)(x, e_n)e_n = \sum_{n=1}^{k} \varphi(\lambda_n)(x, e_n)e_n + c\sum_{n=1}^{k}(x, e_n)e_n$$
$$\to \varphi(T)x + cIx,$$

for $k \to \infty$, we see that $f(T)x = \varphi(T)x + cIx$ is well defined for all $x \in H$. (Notice that if $c \neq 0$, then the sequence of operators $(f(T_k))$ defined by

$$f(T_k)x = \sum_{n=1}^{k} f(\lambda_n)(x, e_n)e_n$$

will *not* converge to $f(T)$ in $B(H)$, since this would imply that $f(T)$ was compact, which is not the case when $c \neq 0$.)

The last statement is trivial since $f(T)^*x = \sum_{n=1}^{\infty} \overline{f}(\lambda_n)(x, e_n)e_n$. ∎

REMARK 6.1 The proposition above is sometimes formulated as

$$\sigma(f(T)) = f(\sigma(T)) \tag{6.6}$$

when f and T satisfy the assumptions. This is a very nice structural statement, and it can be generalized to *normal* bounded operators, that is, operators satisfying $T^*T = TT^*$. ∎

The next application of the Spectral Theorem is to *positive* operators:

DEFINITION 6.3 *An operator T on a Hilbert space is called* positive *if*

$$(Tx, x) \geq 0 \quad \text{for} \quad x \in D(T). \tag{6.7}$$

Notice that a positive and bounded operator is self-adjoint since $((T - T^*)x, x) = 0$ for all $x \in H$ whenever (Tx, x) is real for all $x \in H$. Notice also that all the possible eigenvalues for a positive operator are positive.

Example 6.5
Let T be a compact and positive operator on a Hilbert space H, and let (e_n) be an orthonormal basis for H consisting of eigenvectors for T, with corresponding (positive) eigenvalues (λ_n). We define the operator $\sqrt{T}$ by

$$\sqrt{T}x = \sum_{n=1}^{\infty} \sqrt{(\lambda_n)}(x, e_n)e_n. \tag{6.8}$$

This is again a compact and self-adjoint (and positive) operator according to the proposition. Let us show that $(\sqrt{T})^2 = T$. We have for all $x \in H$ that

$$(\sqrt{T})^2 x = \sum_{n=1}^{\infty} \sqrt{(\lambda_n)}(\sqrt{T}x, e_n)e_n$$

$$= \sum_{n=1}^{\infty} \sqrt{(\lambda_n)}(\sum_{k=1}^{\infty} \sqrt{(\lambda_k)}(x, e_k)e_k, e_n)e_n$$

$$= \sum_{n=1}^{\infty} (\sqrt{(\lambda_n)})^2 (x, e_n)e_n$$

$$= Tx,$$

so $(\sqrt{T})^2 = T$. ▯

Example 6.6

Let T be a compact and self-adjoint operator on a Hilbert space H, and let (e_n) be an orthonormal basis for H consisting of eigenvectors for T, with corresponding eigenvalues (λ_n). Let us make a disjoint partition of the spectrum for T such that

$$\sigma(T) = \{\lambda_n\} = \{\lambda_{n_1}\} \cup \{\lambda_{n_2}\},$$

and let us denote

$$\sigma_1 = \{\lambda_{n_1}\} \quad \text{and} \quad \sigma_2 = \{\lambda_{n_2}\}.$$

Let us define the functions χ_1 and χ_2 on $\sigma(T)$ in the following way:

$$\chi_1(\lambda) = 1 \quad \text{for} \quad \lambda \in \sigma_1$$
$$\chi_1(\lambda) = 0 \quad \text{for} \quad \lambda \in \sigma_2$$
$$\chi_2(\lambda) = 1 \quad \text{for} \quad \lambda \in \sigma_2$$
$$\chi_2(\lambda) = 0 \quad \text{for} \quad \lambda \in \sigma_1.$$

The functions χ_1 and χ_2 are the *characteristic functions* for the sets σ_1 and σ_2. The operators $\chi_1(T)$ and $\chi_2(T)$ are

$$\chi_1(T)x = \sum_{n=1}^{\infty} \chi_1(\lambda_n)(x, e_n)e_n = \sum_{n_1}(x, e_{n_1})e_{n_1} \quad \text{and}$$

$$\chi_2(T)x = \sum_{n=1}^{\infty} \chi_2(\lambda_n)(x, e_n)e_n = \sum_{n_2}(x, e_{n_2})e_{n_2},$$

that is, the orthogonal projections on the closed subspaces spanned by the (e_{n_1}) and the (e_{n_2}), respectively. Since these subspaces are orthogonal, we have that

$$\chi_1(T)\chi_2(T) = 0,$$

and since (e_n) is a basis, we see that

$$\chi_1(T) + \chi_2(T) = I.$$

Thus we see that a decomposition of the spectrum corresponds to a certain orthogonal decomposition of the Hilbert space given by

$$H = \chi_1(T)(H) \oplus \chi_2(T)(H), \tag{6.9}$$

a fact that is very important in stability analysis and control engineering.
$\square$

Now let T be an unbounded and self-adjoint operator. Then we know that $T = T^*$ is closed and $D(T)$ is dense in H. Then the operator $(T - \lambda I)$

is also densely defined and self-adjoint for $\lambda \in R$. Since T is self-adjoint, we know that $\sigma(T) \subset R$, so for $\lambda \in R \setminus \sigma(T)$ the resolvent $R_\lambda(T)$ is a densely defined, self-adjoint, and bounded operator. Let us assume that $R_\lambda(T)$ is also compact. Then, for a fixed λ, we have an orthonormal basis of eigenvectors (e_n) for $R_\lambda(T)$ with corresponding eigenvalues (μ_n) converging to 0. Since

$$R_\lambda(T)e_n = \mu_n e_n$$

for all $n \in N$, we see that $e_n \in D(T)$ and

$$e_n = \mu_n(T - \lambda I)e_n,$$

so for $\mu_n \neq 0$ we find that

$$\frac{1}{\mu_n} = (T - \lambda I)e_n,$$

showing that (e_n) is an orthonormal basis for H consisting of eigenvectors for the unbounded operator $(T - \lambda I)$, with corresponding eigenvalues $(\frac{1}{\mu_n})$ where $|\frac{1}{\mu_n}| \to \infty$. We see that if $0 \notin \sigma(T)$ all this holds with $\lambda = 0$, so for such T we have easily a spectral theorem:

$$Tx = \sum_{n=1}^{\infty} \frac{1}{\mu_n}(x, e_n)e_n \quad \text{for all} \quad x \in D(T). \tag{6.10}$$

Chapter 7

Integral Operators

Perhaps the most common problems in applied mathematics are various kinds of differential equations, since the classical formulations from physics or chemistry are of this nature. The ways of solving these equations are of course many, but one very common method is to reformulate the differential equation into an integral equation. Or, equivalently, instead of formulating the problem as a differential equation, one could from the very start formulate it as an integral equation. The first operators to be considered (apart from matrices) were the integral operators stemming from integral equations of mathematical physics, and it is not wrong to say that the study of integral operators gave birth to modern functional analysis.

7.1   Introduction

In this book we will only consider the operators with kernel $k(x,t)$ depending on two variables, but most of what we will do applies, with minor changes, to higher dimensions. So the results for the simplest integral operators can immediately be translated into more complicated situations.

DEFINITION 7.1

Let k be a function of two variables $(x,t) \in I \times I = I^2$ where I is a finite or infinite real interval. We define a linear integral operator K with kernel $k(x,t)$ as

$$Ku(x) = \int_I k(x,t)u(t)dt, \quad x \in I,$$

whenever this integral makes sense. The domain $D(K)$ will have to be specified in order to accomplish this.

The main objects of our study will of course be integral operators on

$L^2(I)$, and as mentioned in Chapter 2, there is a problem with the point-wise definition of functions in $L^2(I)$ that are not (piecewise) continuous. Therefore, the correct way of understanding a statement like $k(\cdot,t) \in L^2(I)$ is that the function $x \to k(x,t)$ is in $L^2(I)$ for all $t \in I \setminus \Theta$, where Θ is a so-called *Lebesgue nullset*. This is a set with the (somewhat tricky) property that there exists a set $\tilde{\Theta}$ such that $\Theta \subset \tilde{\Theta}$ and $\int_{\tilde{\Theta}} dx$ is well defined and has the value 0. An example of a nullset in R is the rational numbers Q and all subsets thereof. In order to understand this properly, one has to consult a book on *measure theory*, t.ex. [26], but this is really not necessary for the reading of this book. This is the tradeoff we must accept when dealing with L^2-spaces, but fortunately we can allways rely on the densesness of the continuous functions.

7.2   The Class of Hilbert-Schmidt Operators

Many of the integral operators encountered in applications are bounded operators, and many of these are, in fact, compact operators. Examples of this are the differential operators with compact resolvents, since the resolvents often can be written as integral operators. The most important class of compact operators are the *Hilbert-Schmidt operators*.

DEFINITION 7.2
An integral operator on $L^2(I)$ is called a Hilbert-Schmidt *operator if the kernel k is in $L^2(I \times I)$, that is if*

$$\|k\|^2 = \int_I \int_I |k(x,t)|^2 dx dt < \infty. \tag{7.1}$$

One can show that a Hilbert-Schmidt operator is uniqely determined by its kernel, that is, if

$$\int_I k(x,t)u(t)dt = \int_I h(x,t)u(t)dt,$$

for all $u \in L^2(I)$, then $k = h$ in $L^2(I \times I)$.

PROPOSITION 7.1
A Hilbert-Schmidt operator is bounded and

$$\|K\| \le \|k\|. \tag{7.2}$$

PROOF Assume that K is a Hilbert-Schmidt operator with kernel $k \in L^2(I \times I)$. Let $u \in L^2(I)$, then from *Fubini's theorem* in measure theory, the function $x \to \int_I |k(x,t)|^2 dt$ is in $L^1(I)$ such that $\int_I \int_I |k(x,t)|^2 dx dt < \infty$. Thus, it is allowed to use the Cauchy-Schwarz inequality such that

$$|Ku(x)| = |\int_I k(x,t)u(t)dt| \leq \|k(x,\cdot)\|\|u\|$$

since $\|\bar{u}\| = \|u\|$. Hence

$$\int_I |Ku(x)|^2 dx \leq \int_I \|k(x,\cdot)\|^2 \|u\|^2 dx$$
$$= \int_I \|k(x,\cdot)\|^2 dx \|u\|^2$$
$$= \int_I \int_I |k(x,t)|^2 dt dx \|u\|^2$$
$$= \|k\|^2 \|u\|^2,$$

and we see that $\|Ku\| \leq \|k\|\|u\|$. ∎

In order to prove that a Hilbert-Schmidt operator is compact, we must construct a basis for $L^2(I \times I)$ from a basis for $L^2(I)$. This is done by the so-called *tensor product*, which is defined in the following manner: let $f, g \in L^2(I)$. Then we define the function $f \otimes g$ by

$$f \otimes g(x,y) = f(x)g(y), \quad (x,y) \in I \times I. \tag{7.3}$$

It is obvious that $f \otimes g \in L^2(I \times I)$ since

$$\int_I \int_I |f \otimes g(x,y)|^2 dx dy = \int_I |f(x)|^2 dx \int_I |g(y)|^2 dy < \infty.$$

PROPOSITION 7.2

Let (e_n) denote an orthonormal basis for $L^2(I)$. Then $(e_n \otimes \overline{e_m})$ is an orthonormal basis for $L^2(I \times I)$.

PROOF
It is obvious that

$$(e_n \otimes \overline{e_m}, e_{n'} \otimes \overline{e_{m'}}) = \int_I \int_I e_n(x)\overline{e_m(t)}\overline{e_{n'}(x)}\overline{\overline{e_{m'}(t)}} dx dt$$
$$= \int_I e_n(x)\overline{e_{n'}(x)}dx \int_I e_m(t)\overline{e_{m'}(t)}dt$$
$$= (e_n, e_{n'})(e_m, e_{m'})$$
$$= \delta_{nn'}\delta_{mm'},$$

so $(e_n \otimes \overline{e_m})$ is an orthonormal system. To see that it is also a basis, assume that $k \in L^2(I \times I)$ and

$$(k, e_n \otimes \overline{e_m}) = 0, \quad n, m \in N.$$

Then, for all $n, m \in N$ we have

$$
\begin{aligned}
0 &= \int_I \int_I k(x,t)\overline{e_n(x)\overline{e_m(t)}}dtdx \\
&= \int_I (\int_I k(x,t)e_m(t)dt)\overline{e_n(x)}dx \\
&= \int_I Ke_m(x)\overline{e_n(x)}dx \\
&= (Ke_m, e_n),
\end{aligned}
$$

where K is the integral operator with kernel k. Since (e_n) is an orthonormal basis, this shows us that $Ke_m = 0$ for all $m \in N$, so K is the zero operator. Since a Hilbert-Schmidt operator is uniquely determined by its kernel, we see that $k = 0$ in $L^2(I \times I)$, so the sequence $(e_n \otimes \overline{e_m})$ is total. Hence it is an orthonormal basis. ∎

PROPOSITION 7.3

Every Hilbert-Schmidt operator is compact.

PROOF

Let $k \in L^2(I \times I)$ be the kernel for the Hilbert-Schmidt operator K. Then

$$k = \sum_{n=1}^{\infty} \sum_{m=1}^{\infty} (k, e_n \otimes \overline{e_m})e_n \otimes \overline{e_m}, \tag{7.4}$$

and we see that since $\|K\| \leq \|k\|$, we can approximate K in operator norm by the finite rank operators K_{NM} with kernels k_{NM} given by

$$k_{NM} = \sum_{n=1}^{N} \sum_{m=1}^{M} (k, e_n \otimes \overline{e_m})e_n \otimes \overline{e_m},$$

Hence K is compact. ∎

We actually have the following nice characterization of Hilbert-Schmidt operators.

PROPOSITION 7.4

If K is a Hilbert-Schmidt operator on $L^2(I)$ and (e_n) is an orthonormal set in $L^2(I)$, then $\sum_{n=1}^{\infty} \|Ke_n\|^2 < \infty$.

If, on the other hand, $K \in B(L^2(I))$ is such that $\sum_{n=1}^{\infty} \|Ke_n\|^2 < \infty$ for some orthonormal basis (e_n) in $L^2(I)$, then K is a Hilbert-Schmidt operator.

PROOF

Assume that (e_n) is an orthonormal basis for $L^2(I)$ and take $(e_n \otimes \overline{e_m})$ as an orthonormal basis for $L^2(I \times I)$. If K is bounded, we have from Parseval's equation that

$$\sum_{m=1}^{\infty} \|Ke_m\|^2 = \sum_{m=1}^{\infty} \sum_{n=1}^{\infty} |(Ke_m, e_n)|^2.$$

But if $k \in L^2(I \times I)$, we have from the calculations above that

$$(k, e_n \otimes \overline{e_m}) = (Ke_m, e_n),$$

and we see that

$$\sum_{m=1}^{\infty} \|Ke_m\|^2 = \sum_{m=1}^{\infty} \sum_{n=1}^{\infty} |(k, e_n \otimes \overline{e_m})|^2$$
$$= \|k\|^2$$
$$< \infty.$$

If, on the other hand, $\sum_{m=1}^{\infty} \|Ke_m\|^2$ is finite, then so is

$$\sum_{m=1}^{\infty} \sum_{n=1}^{\infty} |(k, e_n \otimes \overline{e_m})|^2,$$

hence

$$\sum_{m=1}^{\infty} \sum_{n=1}^{\infty} (Ke_m, e_n) e_n \otimes \overline{e_m}$$

converges in $L^2(I \times I)$ to some $k \in L^2(I \times I)$ with Fourier coefficients

$$(k, e_n \otimes \overline{e_m}) = (Ke_m, e_n),$$

so K has the kernel k and is a Hilbert-Schmidt operator. ∎

In Example 5.2 we saw that the adjoint of $Ku(x) = \int_I k(x, y)u(y)dy$ in $B(L^2(I))$ was the operator K^* given by

$$K^*u(x) = \int_I \overline{k(y, x)}u(y)dy,$$

so any integral operator in $B(L^2(I))$ is self-adjoint if the kernel satisfies $k(x,y) = \overline{k(y,x)}$. A function satisfying this is said to be *hermitian*. We thus see that Hilbert-Schmidt operators with hermitian kernels are compact self-adjoint operators, and we can formulate a special version of the spectral theorem.

THEOREM 7.5

Assume that K is a Hilbert-Schmidt operator on $L^2(I)$ with hermitian kernel. Then there is an orthonormal basis (e_n) for $L^2(I)$ consisting of eigenfunctions for K with associated real eigenvalues (λ_n). The sequence (λ_n) can be arranged such that $|\lambda_1| \geq |\lambda_2| \geq ...$, where $\lambda_n \to 0$ for $n \to \infty$, and we have that

$$Ku = \sum_{n=1}^{\infty} \lambda_n (u, e_n)e_n, \quad u \in L^2(I).$$

The spectrum of K consists of the eigenvalues (λ_n) together with $\{0\}$, with the possibility of 0 being an eigenvalue or in the continuous spectrum for K.

REMARK 7.1

If $\lambda_n = 0$ from a certain step, 0 is an eigenvalue and, of course, we still have that $\lambda_n \to 0$ for $n \to \infty$. ∎

Example 7.1

Let $H = L^2([0;1])$ and consider the integral operator K with kernel $k(x,t) = xt$. Obviously the kernel is hermitian, so K is self-adjoint. Since for $u \in H$

$$Ku(x) = \int_0^1 xtu(t)dt = \alpha x,$$

for some $\alpha \in C$, we see that K has rank 1 and $K(H)$ is the one-dimensional subspace of H spanned by the function $id(x) = x$. Since

$$\|id\|^2 = \int_0^1 x^2 dx = \frac{1}{3},$$

we see that $\varphi_1(x) = \sqrt{3}x$ is a normalized eigenfunction. Applying K

gives us

$$K\varphi_1(x) = \int_0^1 t\varphi_1(t)dt \cdot x$$

$$= \int_0^1 t\varphi_1(t)dt \cdot \frac{1}{\sqrt{3}}\varphi_1(x)$$

$$= \int_0^1 t^2 dt \varphi_1(x)$$

$$= \frac{1}{3}\varphi_1(x),$$

so $\lambda = \frac{1}{3}$ is the only nonzero eigenvalue. We conclude that

$$\sigma(K) = \sigma_p(K) = \{0; \frac{1}{3}\},$$

so we can write

$$Ku = \frac{1}{3}(u, \varphi_1)\varphi_1, \quad u \in H.$$

Recalling that one of the numbers $\pm\|K\|$ must be an eigenvalue, we conclude that

$$\|K\| = \frac{1}{3}.$$

⬜

From the preceeding propositions we have

COROLLARY 7.6

A bounded linear operator T on $L^2(I)$, where I is an interval, is a Hilbert-Schmidt operator if and only if it is compact and the sum of the squares of the eigenvalues of $\sqrt{T^*T}$ is finite.

PROOF

If (λ_j) is a sequence of scalars and (e_j) an orthonormal basis for $L^2(I)$ such that

$$\sqrt{T^*T}e_j = \lambda_j e_j, \tag{7.5}$$

then

$$\sum_{j=1}^{\infty}\lambda_j^2 = \sum_{j=1}^{\infty}(T^*Te_j, e_j) = \sum_{j=1}^{\infty}\|Te_j\|^2. \tag{7.6}$$

If T is a Hilbert-Schmidt operator, then since $\sqrt{T^*T}$ is compact and self-adjoint, it has an orthonormal basis consisting of eigenvectors, and the above together with the preceeding proposition implies that the sum of the squares of the eigenvalues is finite. If, on the other hand, $\sqrt{T^*T}$ is compact and self-adjoint, then the same arguments imply that T is a Hilbert-Schmidt operator. ∎

7.3   Integral Equations

As mentioned in the introduction, some of the first applications of operator theory were in the study of integral operators and the integral equations that they come from. We will here briefly discuss the so-called *Fredholm* and *Volterra* integral equations that occur frequently in problems related to mathematical physics. We will assume in the following that K is a Hilbert-Schmidt operator on $L^2(I)$ where I is a bounded interval. The kernel for K will be denoted $k(x,t)$, and we have that $k \in L^2(I \times I)$.

DEFINITION 7.3   *A Fredholm integral equation of the first kind is an integral equation of the form*

$$Ku = f, \quad f \in L^2(I). \tag{7.7}$$

A Fredholm integral equation of the second kind is an integral equation of the form

$$(K - \lambda I)u = f, \quad f \in L^2(I) \quad \lambda \neq 0. \tag{7.8}$$

We have met the Fredholm integral equation of the second kind before as a special case of Theorem 6.10. Notice that putting $\lambda = 0$ in an equation of the second kind gives an equation of the first kind, but it is convenient to distinguish between the two. The reason is that the solution operator K^{-1} to the equation of the first kind (if it exists) must be unbounded since K is compact. Therefore, it is not to be expected that the solution has a nice dependence on the data f. Recall that in applications a solution u to $Ku = f$ will often be found by an approximation procedure: we find an expression for K^{-1} that works on a set of well-behaved functions (f_n) chosen such that $f_n \to f$ in $L^2(I)$; the hope is then that the sequence $(K^{-1}f_n)$ converges to the solution u. This will always be true when K^{-1} is bounded, but when K^{-1} is unbounded we cannot conclude anything in general.

If the kernel $k(x,t)$ has the special property that

$$k(x,t) = 0 \quad \text{for} \quad t > x,$$

we see that if $I = [a; b]$, then

$$\int_a^b k(x,t)u(t)dt = \int_a^x k(x,t)u(t)dt, \tag{7.9}$$

so the independent variable x will occur in the limit of the integration. The corresponding integral equations are called *Volterra equations*.

DEFINITION 7.4

A Volterra integral equation of the first kind *is an integral equation of the form*

$$\int_a^x k(x,t)u(t)dt = f(x), \quad f \in L^2(I). \tag{7.10}$$

A Volterra integral equation of the second kind *is an integral equation of the form*

$$\int_a^x k(x,t)u(t)dt - \lambda u(x) = f(x), \quad f \in L^2(I), \quad \lambda \neq 0. \tag{7.11}$$

REMARK 7.2

Notice that a *hermitian* kernel satisfying $k(x,t) = 0$ for $t > x$ must be identically zero for $t \neq x$, hence it is equivalent to the zero function in $L^2(I \times I)$. So the only self-adjoint Hilbert-Schmidt operator stemming from a Volterra equation is the zero operator! ∎

From Exercise 23 we can actually conclude (try !) the following.

PROPOSITION 7.7

If K is a Hilbert-Schmidt operator stemming from a Volterra equation, then 0 is the only possible eigenvalue.

Example 7.2

Consider the Fredholm integral equation of the second kind

$$(K - \lambda I)u = f, \quad f \in L^2(I), \quad \lambda \neq 0.$$

If $|\lambda| < \|K\|^{-1}$ we write the equation as

$$-\lambda(I - \frac{1}{\lambda}K)u = f,$$

and we have the solution

$$u = \frac{-1}{\lambda}(I - \frac{1}{\lambda}K)^{-1}f \in L^2(I).$$

Moreover,

$$(I - \frac{1}{\lambda}K)^{-1} = \sum_{n=0}^{\infty} \frac{1}{\lambda^n}K^n,$$

so we can write explicitly

$$u = \frac{-1}{\lambda}\sum_{n=0}^{\infty}\frac{1}{\lambda^n}K^nf.$$

It is easy to see that the kernel k_n for the operator K^n can be calculated recursively by

$$k_n(x,t) = \int_a^b k(x,s)k_{n-1}(s,t)ds;$$

these are the so-called *iterated kernels*.

⬜

Example 7.3

If the Hilbert-Schmidt operator K from the preceeding example is self-adjoint, we see that we have already solved the equation $(K - \lambda I)u = f$ by the Fredholm alternative (see Theorem 6.10 and Example 6.4). The solution is

$$u = \sum_{i=1}^{\infty}(\lambda_i - \lambda)^{-1}(f, e_i)e_i, \tag{7.12}$$

where it is assumed that λ is not 0 or an eigenvalue, and (e_i) is an orthonormal basis of eigenvectors with corresponding eigenvalues (λ_i).   ⬜

Chapter 8

Semigroups of Evolution

In this chapter we will study how the solution formula

$$x(t) = e^{At}x_0 + \int_0^t e^{A(t-s)}f(s)ds$$

to the ordinary differential equation

$$\frac{d}{dt}x(t) = Ax(t) + f(t),$$

$$x(0) = x_0 \in R^n,$$

where x and f denote sufficiently smooth functions with values in R^n and A is an $n \times n$ matrix, can be generalized to give a solution formula for more general equations. The extension to bounded operators A is not complicated, but in order to consider also partial differential equations of the above type, a more elaborate theory is necessary.

8.1 Strongly Continuous Semigroups

In order to understand and justify what will happen, let us review the finite dimensional case. Consider the equation

$$\frac{d}{dt}x(t) = Ax(t) \quad \text{for } t > 0,$$

$$x(0) = x_0 \in R^n$$

where $A \in B(R^n)$, so A is just an $n \times n$ matrix. The exponential matrix

$$e^{At} = \sum_{k=0}^{\infty} \frac{(At)^k}{k!}$$

is well defined, and $e^{At} \in B(R^n)$ for all $t \in R$ since the series converges absolutely in the Banach space $B(R^n)$. Then, for all t:

$$\frac{d}{dt}e^{At}x_0 = \lim_{h \to 0} \frac{1}{h}(e^{A(t+h)}x_0 - e^{At}x_0)$$

$$= \lim_{h \to 0} \left(\lim_{N \to \infty} \left(\sum_{k=0}^{N} \frac{(t+h)^k - t^k}{k!} A^k \right) x_0 \right)$$

$$= \lim_{N \to \infty} \left(\sum_{k=1}^{N} \frac{t^{k-1}}{(k-1)!} A^k \right) x_0$$

$$= \lim_{N \to \infty} A \sum_{k=0}^{N-1} \frac{(At)^k}{k!} x_0$$

$$= Ae^{At}x_0,$$

and we see that $x(t) = e^{At}x_0$ is the solution to the equation.

Notice that the essential properties of the function $S : R \to B(V)$ given by $S(t) = e^{At}$ we used in the calculations were

$$S(0) = I$$
$$S(t+h) = S(t) + S(h),$$

together with the fact that the mapping $t \to S(t)x_0$ from R into V was differentiable for all $x_0 \in V$. If we allow A to be a bounded operator on a Banach space V, exactly the same calculation will show that $e^{At}x_0$ solves the equation

$$\frac{d}{dt}x(t) = Ax(t),$$
$$x(0) = x_0 \in V.$$

Then it is straightforward to represent the solution to the equation

$$\frac{d}{dt}x(t) = Ax(t) + f(t),$$
$$x(0) = x_0 \in V,$$

where $f : R \to V$ is continuous, by the *variation of parameter formula*

$$x(t) = e^{At}x_0 + \int_0^t e^{A(t-s)} f(s)ds.$$

In order to represent also solutions to partial differential equations by an analogue formula it is necessary to extend the theory further, so we introduce the following

DEFINITION 8.1 *Let V be a Banach space. By a* semigroup of bounded linear operators *on V, we understand a family*

$$S(t), 0 \le t < \infty$$

of bounded operators on V satisfying

$$S(0) = I$$
$$S(t + s) = S(t)S(s).$$

The fact that $S(t + s) = S(t)S(s)$ is called *the semigroup property*. Usually, the semigroups encountered in applications have a desirable property called *strong continuity*.

DEFINITION 8.2 *Let $S(t)$, $0 \le t < \infty$ be a semigroup of bounded linear operators on a Banach space V. We say that $S(t)$ is* strongly continuous *or C_0 if, for all $x \in V$, we have that*

$$S(t)x \to x \quad \text{for} \quad t \to 0^+.$$

Notice that strong continuity of $S(t)$ does not imply that $S(t) \to I$ in $B(V)$ for $t \to 0$; this is a much stronger statement. If this is also the case, the semigroup is called *uniformly continuous at* 0. Notice also that if $S(t)$ is a C_0-semigroup, then

$$S(t)x \to S(s)x \quad \text{for all} \quad x \in V, \quad \text{and} \quad t \to s^+.$$

This is an immediate consequence of the semigroup property.

Since the idea is to use the semigroup to represent the solution to equations of the above type where the operator A is unbounded, we see that necessarily the semigroup must be related to the operator by the formula

$$\frac{d}{dt}S(t)x = AS(t)x$$

for all the values of $x \in V$ that we would like to use as initial conditions in the equation. This is the motivation for the following.

DEFINITION 8.3 *Let $S(t)$, $0 \le t < \infty$ be a semigroup of bounded linear operators on a Banach space V. The* infinitesimal generator A *of the semigroup is the operator defined by*

$$Ax = \lim_{t \to 0^+} \frac{1}{t}(S(t)x - x), \tag{8.1}$$

whenever the limit exists. The domain of the infinitesimal generator is defined as

$$D(A) = \{x \in V \mid Ax \in V\} \tag{8.2}$$

It is easy to see that A is a linear operator, and usually we write $S(t) = e^{At}$ whenever $S(t)$ is generated by A.

REMARK 8.1   It is not difficult to show that the semigroup $S(t)$ is uniformly continuous at 0 if and only if A is a bounded operator. This is why these occur rarely in applications.   ▮

Since $S(t)$ is a bounded operator for any $t \geq 0$, we have that $\|S(t)x\| \leq \|S(t)\|\|x\|$, but this contains no information of how $\|S(t)x\|$ develops, unless we know how $\|S(t)\|$ behaves. The next theorem states that $\|S(t)\|$ grows at most in an exponential manner if it is a C_0-semigroup.

THEOREM 8.1
Let $S(t)$, $0 \leq t < \infty$ be a C_0-semigroup on a Banach space V. Then there exists $\alpha \in R$ and a constant $M \geq 1$ such that

$$\|S(t)\| \leq Me^{\alpha t} \tag{8.3}$$

for all $t \geq 0$.

PROOF   Let $T > 0$ be given and define

$$M = \sup_{0 \leq t \leq T} \{\|S(t)\|\}.$$

Then $M \geq 1$ since $S(0) = I$. We will show that M is finite, so assume that $M = \infty$. Then there is a sequence (t_n) in the interval $[0; T]$ for which

$$\|S(t_n)\| \to \infty \quad \text{for} \quad n \to \infty,$$

and by the Principle of Uniform Boundedness there is an $x \in V$ such that

$$\|S(t_n)x\| \to \infty \quad \text{for} \quad n \to \infty.$$

But since (t_n) is a sequence in the compact interval $[0; T]$, it has a convergent subsequence (t_{n_k}), $t_{n_k} \to t_0$ for $k \to \infty$. But then

$$\|S(t_0)x\| = \lim_k \|S(t_{n_k})x\| = \infty,$$

which is a contradiction, so M is finite. But any $t > 0$ can be written $t = nT + r(t)$ where $0 \leq r(t) < T$ and $n \in N_0$, so

$$S(t) = S(nT + r(t)) = S(T)^n S(r(t)).$$

Now, let $\alpha = \frac{\log(M)}{T}$, then

$$
\begin{aligned}
\|S(t)\| &= \|S(T)^n S(r(t))\| \\
&\leq \|S(T)\|^n \|S(r(t))\| \\
&\leq M^n M \\
&= e^{n\log(M)} M \\
&= e^{\frac{t-r(t)}{T}\log(M)} M \\
&= e^{\alpha t} e^{\frac{-r(t)\log(M)}{T}} M \\
&\leq M e^{\alpha t},
\end{aligned}
$$

since $\log(M) \geq 0$. ■

REMARK 8.2

If A is an unbounded operator on a Hilbert space H with a sequence of eigenvalues (λ_n) where $Re(\lambda_n) \to \infty$, then A is not the generator of a C_0-semigroup. To see this, we notice that if x_n is an eigenvector, that is,

$$Ax_n = \lambda_n x_n,$$

then if A generates a C_0-semigroup $S(t)$

$$S(t)x_n = e^{\lambda_n t} x_n$$

since $\frac{d}{dt} S(t)x_n = Ax_n$ at $t = 0$. But

$$\|S(t)x_n\| = e^{Re(\lambda_n)t} \|x_n\|$$

such that $\|S(t)\|$ cannot be bounded by $Me^{\alpha t}$ when $Re(\lambda_n) \to \infty$. ■

It is a highly desirable feature in control engineering when the growth constant α above is negative, reflecting the fact that the energy of the system decreases exponentially for all initial configurations. This justifies the following definition.

DEFINITION 8.4

We say that a C_0-semigroup is exponentially stable *or* decays exponentially *if there exist $M \geq 1$ and $\omega > 0$ such that*

$$\|S(t)\| \leq Me^{-\omega t} \tag{8.4}$$

for all $t \geq 0$.

It is usually hard to determine ω above explicitly, so the following theorem is often useful:

THEOREM 8.2
A C_0-semigroup decays exponentially if and only if there is a $T > 0$ such that
$$\|S(T)\| < 1.$$

PROOF   If $\|S(t)\| \leq Me^{-\omega t}$, $M \geq 1$, $\omega > 0$, take $T = \omega^{-1} \log(2M)$ and we find that $\|S(t)\| \leq \frac{1}{2}$.

On the other hand, if $\|S(T)\| = \rho < 1$, then we can write $t \geq 0$ as $t = nT + r(t)$ where $0 \leq r(t) < T$ and get

$$\begin{aligned}
\|S(t)\| &= \|S(nT)S(r(t))\| \\
&\leq \|S(T)\|^n \|S(r(t))\| \\
&\leq \rho^n M
\end{aligned}$$

where $M = \sup_{0 \leq t \leq T} \|S(t)\|$. Continuing the calculation we find

$$\begin{aligned}
\|S(t)\| &\leq Me^{n \log(\rho)} \\
&= Me^{\frac{((t-r(t))}{T}) \log(\rho)} \\
&= Me^{\frac{\log(\rho)t}{T}} e^{\frac{-r(t) \log(\rho)}{T}} \\
&\leq M'e^{-\omega t},
\end{aligned}$$

where we have put $\omega = \frac{-\log(\rho)}{T} (> 0)$ and $M' = \frac{M}{\rho}$.   ∎

THEOREM 8.3
Let A be the infinitesimal generator of a C_0-semigroup $S(t)$ on a Banach space V. Then the domain $D(A) = \{x \in V \mid Ax \in V\}$ of A is dense in V, and A is a closed operator.

PROOF   Let $x \in V$ and observe that since $s \to S(s)x$ is continuous from R to V, it is straightforward to define $\int_0^t S(s)x\,ds$ for any fixed $t \geq$ as a Riemann integral. Then we define

$$x_t = \frac{1}{t} \int_0^t S(s)x\,ds, \tag{8.5}$$

and observe that for any $x \in V$, we have that $\|x_t - x\| \to 0$ for $t \to 0^+$. This means that the set

$$\{x_t \mid x \in V, \quad t \geq 0\}$$

is dense in V. Now we show that all the x_t's are contained in $D(A)$, so $D(A)$ must also be dense. For this we must show that the limit

$$\lim_{\epsilon \to 0^+} \frac{S(\epsilon)x_t - x_t}{\epsilon}$$

exists for all $t \geq 0$. So let $t > 0$ and get

$$\begin{aligned}
\frac{S(\epsilon)x_t - x_t}{\epsilon} &= \frac{1}{\epsilon}(S(\epsilon)(\frac{1}{t}\int_0^t S(s)x ds) - \frac{1}{t}\int_0^t S(s)x ds) \\
&= \frac{1}{t}(\frac{1}{\epsilon}(\int_0^t S(s+\epsilon)x ds - \int_0^t S(s)x ds)) \\
&= \frac{1}{t}(\frac{1}{\epsilon}(\int_\epsilon^{t+\epsilon} S(s)x ds - \int_0^t S(s)x ds)) \\
&= \frac{1}{t}(\frac{1}{\epsilon}(\int_t^{t+\epsilon} S(s)x ds - \int_0^\epsilon S(s)x ds)) \\
&\to \frac{1}{t}(S(t)x - x)
\end{aligned}$$

for $\epsilon \to 0^+$. Hence x_t is in $D(A)$, and $D(A)$ is dense in V.

To show that A is closed, assume that (x_n) is a sequence in $D(A)$ such that $x_n \to x \in V$ and $Ax_n \to y \in V$.

For any fixed $n \in N$,

$$\frac{1}{t}(S(t)x_n - x_n) \to Ax_n \quad \text{for} \quad t \to 0^+,$$

so for any $\epsilon > 0$ we can find $t_0 > 0$ such that

$$\|\frac{1}{t}(S(t)x_n - x_n) - Ax_n\| < \epsilon \quad \text{for all} \quad n \in N$$

and $0 < t < t_0$. Now, letting $n \to \infty$, we see that

$$\|\frac{1}{t}(S(t)x - x) - y\| < \epsilon \quad \text{for} \quad 0 < t < t_0,$$

so $\frac{1}{t}(S(t)x - x)$ converges to y for $t \to 0^+$, hence $x \in D(A)$ with $Ax = y$, and A is closed. ∎

We now have from the proof the following theorem.

THEOREM 8.4
For the equation

$$\frac{d}{dt}x(t) = Ax(t), \quad t > 0,$$
$$x(0) = x_0 \in V,$$

where V denotes a Banach space and A is the infinitesimal generator of a C_0-semigroup, we have for $x_0 \in D(A)$ the unique solution

$$x(t) = S(t)x_0 \tag{8.6}$$

satisfying that

$$x \in C^1([0; \infty[; V) \cap C^0([0; \infty[; D(A)), \tag{8.7}$$

where $D(A)$ is equipped with the graph norm

$$\|x\|_{D(A)} = \|x\| + \|Ax\|, \quad x \in D(A). \tag{8.8}$$

8.2   The Resolvent

Now we will explore the connection between the semigroup $S(t)$ and its infinitesimal generator A. More precisely, we will investigate the resolvent $R_\lambda(A)$ of A. Much of what we will do can be put into a more general framework, but we will limit our presentation to the case where $S(t)$ is a C_0-semigroup on a Hilbert space H with infinitesimal generator A. We will denote the growth constant of $S(t)$ by α such that

$$\|S(t)\| \leq Me^{\alpha t} \tag{8.9}$$

for all $t \geq 0$. Then, for $Re(\lambda) > \alpha$ we can define a bounded, linear operator $R(\lambda)$ on H by

$$R(\lambda)x = \int_0^\infty e^{-\lambda s} S(s)x\, ds, \tag{8.10}$$

and we will show that $R(\lambda)$ is exactly the resolvent $R_\lambda(A)$ of A. Notice for later use that it is obvious from the definition that

$$\|R(\lambda)\| \to 0$$

for $Re(\lambda) \to \infty$.

First we will show that the range of $R(\lambda)$ is $D(A)$, for every λ with $Re(\lambda) > \alpha$. The first step is to calculate, for $\epsilon > 0$:

$$\begin{aligned}
\frac{1}{\epsilon}(S(\epsilon) - I)R(\lambda)x &= \frac{1}{\epsilon}\int_0^\infty e^{-\lambda s}(S(s+\epsilon)x - S(s)x)ds \\
&= \frac{1}{\epsilon}\left(\int_\epsilon^\infty e^{-\lambda s}e^{\lambda \epsilon}S(s)x\, ds - \int_0^\infty e^{-\lambda s}S(s)x\, ds\right) \\
&= -\frac{1}{\epsilon}\int_0^\epsilon e^{-\lambda s}S(s)x\, ds + \frac{1}{\epsilon}(e^{\lambda \epsilon} - 1)\int_\epsilon^\infty e^{-\lambda s}S(s)x\, ds \\
&\to -x + \lambda R(\lambda)x
\end{aligned}$$

for $\epsilon \to 0$. This shows that $R(\lambda)x$ is in $D(A)$ for all $x \in H$ and that

$$\lambda R(\lambda)x - AR(\lambda)x = x. \tag{8.11}$$

On the other hand, if $x \in D(A)$, then

$$\frac{1}{\epsilon}(S(\epsilon) - I)R(\lambda)x = \int_0^\infty e^{-\lambda s}\frac{1}{\epsilon}(S(s + \epsilon)x - S(s)x)ds$$

where

$$\frac{1}{\epsilon}(S(s + \epsilon)x - S(s)x) = S(s)\frac{1}{\epsilon}(S(\epsilon) - I) \to S(s)Ax$$

for $\epsilon \to 0$.

Since

$$\|\frac{1}{\epsilon}(S(s + \epsilon)x - S(s)x) - S(s)Ax\| = \|S(s)(\frac{1}{\epsilon}(S(\epsilon) - I)x - Ax)\|$$

is bounded above by some constant times $\|S(s)\|$, the integral

$$\int_0^\infty e^{-\lambda s}\frac{1}{\epsilon}(S(s + \epsilon)x - S(s)x)ds$$

will converge to

$$\int_0^\infty e^{-\lambda s}S(s)Axds, \tag{8.12}$$

which is exactly $R(\lambda)Ax$. So on $D(A)$, $R(\lambda)$ and A commute, so

$$AR(\lambda)x = R(\lambda)Ax \quad \text{for} \quad x \in D(A),$$

hence from the previous calculation

$$\lambda R(\lambda)x - R(\lambda)Ax = x \quad \text{for} \quad x \in D(A),$$

showing that $D(A)$ is contained in the range of $R(\lambda)$ since

$$R(\lambda)(\lambda x - Ax) = x \quad \text{for} \quad x \in D(A).$$

Since A generates a strongly continuous semigroup, A is closed and $D(A)$ is dense in H, and we have shown that

$$R(\lambda)(\lambda I - A) = (\lambda I - A)R(\lambda) = I$$

on $D(A)$. Hence $R(\lambda) = -R_\lambda(A) = -(A - \lambda I)^{-1}$, and this was what we wanted to show. (The sign of the resolvent in the equation is, of course, depending on the definition of the resolvent.)

We will list some more properties of the resolvent. We now know that if $S(t)$ is a C_0-semigroup with the infinitesimal generator A and growth constant α such that $\|S(t)\| \leq Me^{\alpha t}$ for $t \geq 0$, then for λ with $Re(\lambda) > \alpha$ the resolvent of A is

$$R_\lambda(A)x = -\int_0^\infty e^{-\lambda s} S(s)x\, ds$$

for all $x \in H$.

Moreover, from the equation

$$\lambda R_\lambda(A)x - AR_\lambda(A)x = x$$

we find, since $AR_\lambda(A)x \to 0$ for $Re(\lambda) \to \infty$, and $\|\lambda R_\lambda(A)\|$ is bounded for $Re(\lambda)$ large, that

$$\lambda R_\lambda(A)x \to x \quad \text{for} \quad Re(\lambda) \to \infty.$$

Multiplying the same equation with λ gives

$$\lambda^2 R_\lambda(A)x - \lambda AR_\lambda(A)x = \lambda x,$$

and since for $x \in D(A)$:

$$AR_\lambda(A)x = R_\lambda(A)Ax,$$

we can obtain the limit

$$\lambda^2 R_\lambda(A)x - \lambda x \to Ax \quad \text{for} \quad Re(\lambda) \to \infty \quad \text{and} \quad x \in D(A).$$

This represents the (unbounded) operator A as a limit of bounded operators and is called *the Yosida approximation* of A.

Notice also that we have *the resolvent equation*:

$$R_\lambda(A) - R_\mu(A) = (\mu - \lambda)R_\mu(A)R_\lambda(A),$$

which is straightforward to calculate.

Chapter 9

Sobolev Spaces

The analysis of PDE's naturally involves function spaces that are not only defined in terms of the properties of the function itself, but also in terms of the properties of its derivatives. Sobolev spaces prove to be useful tools in this analysis. For a comprehensive study we refer to Adams [1] . Sobolev spaces are Banach spaces by construction, and the new feature is that in order for a function to belong to a certain Sobolev space, both the function itself and its derivatives, up to a certain order, must lie in a certain L^p-space. Since we do not attempt to give anything but a brief introduction to the subject, we will assume that the boundaries of the domains in R^n we consider are all smooth. Sobolev spaces in nonsmooth domains are studied in detail in Grisvard [7].

9.1 Basic Definitions

Let $\Omega \subset R^n$ denote an open set. We will say that a function f is *locally integrable* if

$$\int_K f(x)dx < \infty \qquad (9.1)$$

for all compact sets $K \subset \Omega$. If this is the case, we write

$$f \in L^1_{loc}(\Omega).$$

Assume that f is locally integrable and observe that for all $\varphi \in C_0^\infty(\Omega)$, the integral $\int_\Omega f\varphi dx$ exists. If, moreover, $f \in C^1(\Omega)$, we see that

$$\int_\Omega f'\varphi dx = - \int_\Omega f\varphi' dx \qquad (9.2)$$

since the support of φ is compact. Repeating this argument we find that

$$\int_\Omega f^{(k)}\varphi\,dx = (-1)^k \int_\Omega f\varphi^{(k)}\,dx, \tag{9.3}$$

if $f \in C^k(\Omega)$. The right-hand sides of the integrals above makes sense whether f is differentiable or not, and define linear forms on the space $C_0^\infty(\Omega)$. The topology on $C_0^\infty(\Omega)$ is rather complicated, but for our purposes we define a linear form Λ on $C_0^\infty(\Omega)$ to be *continuous* in the following way:

DEFINITION 9.1

We say that a sequence (φ_n) in $C_0^\infty(\Omega)$ converges to 0 and write $\varphi_n \to 0$ for $n \to \infty$ if there exists a compact set $K \subset \Omega$ containing the support of every φ_n, and the sequences of all the derivatives of (φ_n) converge uniformly to 0.

A linear form $\Lambda : C_0^\infty(\Omega) \to C$ is continuous if

$$\Lambda(\varphi_n) \to 0, \tag{9.4}$$

for all sequences $\varphi_n \to 0$ for $n \to \infty$.

The set of all continuous linear forms on $C_0^\infty(\Omega)$ is denoted $D'(\Omega)$, the space of distributions in Ω.

We thus see that every locally integrable function defines a distribution by

$$\Lambda_f(\varphi) = \int_\Omega f\varphi\,dx, \tag{9.5}$$

and we will use the notation

$$\langle f, \varphi \rangle = \int_\Omega f\varphi\,dx \tag{9.6}$$

for a general distribution f, even when we have no right to write f as a function in the integral.

DEFINITION 9.2   *Let f be a distribution on Ω. The derivative of f, $f^{(k)}$, called the derivative in sense of distributions, is the distribution defined by*

$$\langle f^{(k)}, \varphi \rangle = (-1)^k \langle f, \varphi^{(k)} \rangle \tag{9.7}$$

for all $\varphi \in C_0^\infty(\Omega)$.

REMARK 9.1   The *delta distribution* δ_x is defined by

$$\langle \delta_x, \varphi \rangle = \varphi(x), \tag{9.8}$$

and we see that

$$\langle \delta_x^{(k)}, \varphi \rangle = (-1)^k \varphi^{(k)}(x). \tag{9.9}$$

∎

We choose to define Sobolev spaces in the following manner.

DEFINITION 9.3   *Let Ω be an open subset in R^n. For $1 \leq p \leq \infty$ the Sobolev spaces denoted by $H^{k,p}(\Omega)$ are the spaces of functions $u \in L^p(\Omega)$ such that $D^\alpha u \in L^p(\Omega)$ for all $\alpha = (\alpha_1, .., \alpha_n) \in Z_+^n, \mid \alpha \mid = \alpha_1 + ... + \alpha_n \leq k$.*

Here we use the so-called *multiindex notation*

$$D^\alpha = D_1^{\alpha_1} D_2^{\alpha_2} \cdots D_n^{\alpha_n}$$

where $\alpha = (\alpha_1, .., \alpha_n) \in Z_+^n$ and $D_j = \frac{\partial}{\partial x_j}$. $H^{k,p}$ is equipped with the norm

$$\|u\|_{H^{k,p}(\Omega)} = \left(\sum_{|\alpha| \leq k} \|D^\alpha u\|_{L^p(\Omega)}^p \right)^{\frac{1}{p}} \quad \text{for } 1 \leq p < \infty.$$

$$\tag{9.10}$$

$$\|u\|_{H^{k,\infty}(\Omega)} = \sup_{|\alpha| \leq k} \|D^\alpha u\|_{L^\infty(\Omega)}.$$

REMARK 9.2   The notation $W^{k,p}$ is sometimes used instead of $H^{k,p}$.
∎

REMARK 9.3   In the case where $p = 2$, the Sobolev spaces are denoted $H^k(\Omega)$ and the inner product in $H^k(\Omega)$ is defined by

$$(u, v)_{H^k(\Omega)} = \sum_{|\alpha| \leq k} \int_\Omega D^\alpha u(x) \overline{D^\alpha v(x)} dx.$$

∎

We have the following proposition that we state without proof:

PROPOSITION 9.1

$H^{k,p}(\Omega)$ is a Banach space. If $1 < p < \infty$, it is a reflexive Banach space. If $p = 2$, it is a Hilbert space.

The case $p = \infty$ is not often used because $L^\infty(\Omega)$ has some unpleasant properties. Not only is it not reflexive, which also is the case for $L^1(\Omega)$, but it is not separable, which $L^1(\Omega)$ is.

It is well known that if $p < +\infty$, then $C_0^\infty(\Omega)$ is dense in $L^p(\Omega)$, but this is not generally the case for $H^{k,p}(\Omega)$. Let us point out this important fact.

PROPOSITION 9.2

Let Ω be bounded, let $k \geq 1$ and $1 \leq p \leq +\infty$. Then $C_0^\infty(\Omega)$ is not dense in $H^{k,p}(\Omega)$.

For a proof, see Renardy and Rogers [23], p.206.

It is therefore natural to have the following definition.

DEFINITION 9.4   By $H_0^{k,p}(\Omega)$ we denote the closure of $C_0^\infty(\Omega)$ in $H^{k,p}(\Omega)$. For $p = 2$ it is denoted by $H_0^k(\Omega)$.

REMARK 9.4   The notation $W_0^{k,p}(\Omega)$ is also used instead of $H_0^{k,p}(\Omega)$.

We recall from Definition 9.3 that the functions u of Sobolev spaces are functions where all the derivatives of u of order less than or equal k are functions of $L^p(\Omega)$. We can, however, actually do things a litle more general; we can define Sobolev spaces for $k \in R$. Let us for a moment denote this space by $\widetilde{H}^s(R^n)$ and give the definition.

DEFINITION 9.5   Let $s \in R$. Then we say that a function u in R^n belongs to $\widetilde{H}^s(R^n)$ if the Fourier transform $\hat{u}$ is a square-integrable function with respect to the weight $(1 + |\xi|^2)^s d\xi$.

In Definition 9.5, s can be any real number, but for our purposes the case $s \geq 0$ is sufficient and we will only consider this case.

We can turn $\widetilde{H}^s(R^n)$ into a Hilbert space by introducing the norm

$$\|u\|_{\widetilde{H}^s(R^n)} = \frac{1}{(2\pi)^{\frac{n}{2}}} \left\{ \int_{R^n} |\hat{u}|^2 (1 + |\xi|^2)^s d\xi \right\}^{\frac{1}{2}}. \qquad (9.11)$$

If we Fourier transform $u \in H^k(R^n)$ we get, according to Definition 9.3, that

$$\|u\|_{H^k(R^n)} = \frac{1}{(2\pi)^{\frac{n}{2}}} \left\{ \int_{R^n} |\hat{u}|^2 \sum_{|\alpha| \leq k} |\xi^\alpha|^2 \, d\xi \right\}^{\frac{1}{2}}. \qquad (9.12)$$

From the norms (9.11) and (9.12) it is seen that the two spaces are equal. However, the two norms are not equal. They are merely equivalent.

In the sections to come the following topics will be discussed.

- Are there special cases of Sobolev spaces where C^∞-functions are dense? What kind of C^∞-functions do we have to use?

- Extension theorems: If $f \in H^{k,p}(\Omega)$, does there exist an $F \in H^{k,p}(R^n)$ such that $F|_\Omega = f$?

- Special imbedding relations between Sobolev spaces.

- Trace theorems: Since $C_0^\infty(\Omega)$ is dense in L^p, it is not meaningful to talk about boundary values of arbitrary L^p-functions. However, $C_0^\infty(\Omega)$ is generally not dense in $H^{k,p}(\Omega)$. What can we say about boundary values of functions in Sobolev spaces ?

9.2 Density Theorems

It is obviously of interest to know when C^∞-functions are dense in Sobolev spaces. In this section we will give some of the density results which are of importance.

We start with a proposition.

PROPOSITION 9.3
Let Ω be any open subset of R^n. If $1 \leq p < +\infty$, $C^\infty(\Omega) \cap H^{m,p}(\Omega)$ is dense in $H^{m,p}(\Omega)$.

Hence, any function in $H^{m,p}(\Omega)$ can be approximated by functions whose derivatives exist in the classical sense. So in proofs, everything can be carried out for smooth functions, and hereafter results can be obtained in $H^{m,p}(\Omega)$ by taking limits.

In the proof of the Theorem 9.12 (The Trace Theorem) in R^n, we use the following density result:

PROPOSITION 9.4

Whatever the integer $m \geq 0$, whatever p, $1 \leq p < +\infty$, $C_0^\infty(R^n)$ is dense in $H^{k,p}(R^n)$.

PROOF   See Treves [29], p. 216.   ∎

In many cases one would like to have an approximation by functions which are smooth up to the boundary rather than just in the interior of Ω. This demands some assumptions on the boundary Γ of Ω.

DEFINITION 9.6   [k-extension property] We say that Ω has the k-extension property if there is a bounded linear mapping $E : H^k(\Omega) \rightarrow H^k(R^n)$ such that $Eu|_\Omega = u$ for every $u \in H^k(\Omega)$.

On the other hand, every function in $H^k(R^n)$ is also in $H^k(\Omega)$ by its restriction to Ω.
Now we can state a density result concerning $C^\infty(\overline{\Omega})$ functions which are smooth up to the boundary.

PROPOSITION 9.5

Assume that Ω has the k-extension property. Then functions in $C^\infty(\overline{\Omega})$ with bounded support are dense in $H^k(\Omega)$.

PROOF   See Renardy and Rogers [23], p. 210.   ∎

9.3   Extension Theorems

The question we raise here is whether we are able to extend functions to larger domains. We specially consider extension from Ω to R^n. A simple extension result is stated below.

PROPOSITION 9.6

Let $u \in H_0^k(\Omega)$. Then u can be extended to a function in $H^k(R^n)$ by defining it to be zero outside of $\overline{\Omega}$.

PROOF   Take a sequence $u_n \in C_0^\infty(\Omega)$ which converges to u in $H^k(\Omega)$. Then u_n also converges in $H^k(R^n)$.   ∎

The following theorem deals with a more general situation.

THEOREM 9.7
Assume that Ω is bounded and has a smooth boundary Γ. Then, for any integer with $0 \leq k \leq K$, there exists a continuous linear operator

$$E : H^k(\Omega) \to H^k(R^n)$$

such that $Eu|_\Omega = u$. Moreover, E can be chosen independent of k.

PROOF   See Lions and Magenes [19], Vol. I, p. 38.   ∎

9.4   Imbedding Theorems

In the applications of Sobolev spaces later in this book and in partial differential equations in general, it is important to have inequalities between functions in Sobolev spaces. An application of the concept of continuous imbedding gives us some of these inequalities. Let us first define a continuous imbedding.

DEFINITION 9.7   *[Continuous Imbedding]. Let X, Y be Banach spaces. We then say that X is continuously imbedded in Y if X is a subset of Y and there exists a constant C such that*

$$\|x\|_Y \leq C\|x\|_X \text{ for every } x \in X.$$

The identity operator $I : X \to Y$ is said to be the imbedding operator of X into Y. If X is a dense subspace of Y, we say that I is a continuous dense injection.

Example 9.1
If $p > q$ we have that $L^p(\Omega)$ is continuously imbedded in $L^q(\Omega)$.   □

We have that $H^k(\Omega) \subset H^{k'}(\Omega)$ for nonnegative integers $k > k'$, and that the imbedding is continuous. Let us elaborate:

- $H^k(\Omega)$ is continuously imbedded in $H^{k'}(\Omega)$ for nonnegative integers $k > k'$.

- Let $k > k'$. Then for every $\varphi \in H^k(\Omega)$ we have that $\|\varphi\|_{H^{k'}(\Omega)} \leq C\|\varphi\|_{H^k(\Omega)}$ for some constant $C \leq 1$.

- Let $p > q$. Then $H^{k,p}(\Omega)$ is continuously imbedded in $H^{k,q}(\Omega)$.

Example 9.2
Let Ω have a k-extension property. Then it can be shown that $H^{k,p}(\Omega)$ for $kp > n$ is continuously imbedded in $C_b(\Omega)$. See, for instance, Renardy and Rogers [23], p. 215.   ▯

9.4.1   Example

$H^1(\Omega)$ is continuously imbedded in $L^2(\Omega)$. Hence

$$\|\varphi\|_{L^2(\Omega)} \le C\|\varphi\|_{H^1(\Omega)} \quad \forall \varphi \in H^1(\Omega), \tag{9.13}$$

where $C \le 1$. This is equivalent to

$$\|\varphi\|_{L^2(\Omega)} \le C(\|\varphi\|_{L^2(\Omega)} + \|\nabla\varphi\|_{L^2(\Omega)}).$$

Hence

$$\|\varphi\|_{L^2(\Omega)} \le K\|\nabla\varphi\|_{L^2(\Omega)} \tag{9.14}$$

for some constant K.

Ω need not necessarily be bounded, but only bounded in one direction.

THEOREM 9.8 (Poincare's Inequality)
Let Ω be contained in the strip $|x_1| \le d < +\infty$. Then there is a constant c depending only on k and d such that

$$\|u\|_{H^k(\Omega)} \le c \sum_{|\alpha|=k} \|D^\alpha u\|^2_{L^2(\Omega)},$$

for every $u \in H^k_0(\Omega)$.

PROOF
We will give the proof for $k = 1$; the general case follows by induction. By density, it suffices to consider $u \in C^\infty_0(\Omega)$. An integration by part yields

$$\|u\|^2_{L^2(\Omega)} = \int_\Omega 1 \cdot |u(x)|^2 \, dx = -\int_\Omega x_1 \frac{\partial |u(x)|^2}{\partial x_1} \, dx \le 2d\|u\|_{L^2(\Omega)} \left\| \frac{\partial u}{\partial x_1} \right\|_{L^2(\Omega)}. \tag{9.15}$$

∎

9.4.2 Applications

In the following we present some estimates which are relevant in many applications.

LEMMA 9.9

If $\varphi \in H_0^2(\Omega)$, then there exist constants c_1 and c_2 such that

$$c_1\|\varphi\|_{H_0^2(\Omega)}^2 \leq \|\Delta\varphi\|_{L^2(\Omega)}^2 + \|\varphi\|_{L^2(\Omega)}^2 \leq c_2\|\varphi\|_{H_0^2(\Omega)}^2. \tag{9.16}$$

PROOF   Assume that $\varphi \in C_0^\infty(R^n)$. Using the Theorem of Parseval-Plancherel we get:

$$\|\Delta\varphi\|_{L^2(\Omega)}^2 + \|\varphi\|_{L^2(\Omega)}^2 = (2\pi)^{-n} \int_{R^n} (|\xi|^4 + 1) \, |\hat{\varphi}(\xi)|^2 \, d\xi.$$

For suitable constants $\widetilde{k}_1$ and $\widetilde{k}_2$ we have:

$$\widetilde{k}_1(1 + |\xi|^2)^2 \leq |\xi|^4 + 1 \leq \widetilde{k}_2(1 + |\xi|^2)^2.$$

Hence

$$\widetilde{k}_1(2\pi)^{-n} \int_{R^n} (\ 1 + |\xi|^2)^2 \, |\hat{\varphi}(\xi)|^2 \, d\xi \leq (2\pi)^{-n} \int_{R^n} (1 + |\xi|^4) \, |\hat{\varphi}(\xi)|^2 \, d\xi$$

$$\leq \widetilde{k}_2(2\pi)^{-n} \int_{R^n} (1 + |\xi|^2)^2 \, |\hat{\varphi}(\xi)|^2 \, d\xi. \tag{9.17}$$

(9.17) is equivalent to

$$c_1\|\varphi\|_{H^2(\Omega)}^2 \leq \|\Delta\varphi\|_{L^2(\Omega)}^2 + \|\varphi\|_{L^2(\Omega)}^2 \leq c_2\|\varphi\|_{H^2(\Omega)}^2. \tag{9.18}$$

Since $C_0^\infty(R^n)$ is dense in $H^k(R^n)$, (9.18) is valid for $\varphi \in H_0^2(\Omega)$ using Lemma 9.6. The proof of Lemma 9.9 is completed.   ∎

9.4.3 Example

From Lemma 9.9 we immediately obtain

$$\|\varphi\|_{L^2(\Omega)}^2 \leq \frac{1}{\lambda_0}\|\Delta\varphi\|_{L^2(\Omega)}^2 \quad \forall \varphi \in H_0^2(\Omega) \tag{9.19}$$

for some constant λ_0.

LEMMA 9.10

If $\varphi \in H_0^2(\Omega)$, then there exist constants k_1 and k_2 such that

$$k_1\|\varphi\|_{H_0^2(\Omega)}^2 \leq \|\Delta\varphi\|_{L^2(\Omega)}^2 \leq k_2\|\varphi\|_{H_0^2(\Omega)}^2. \tag{9.20}$$

PROOF   The inequality to the right of (9.20) is obvious. Now from Lemma 9.9 there exists a constant c_1 such that

$$c_1 \|\varphi\|^2_{H^2_0(\Omega)} \leq \|\Delta\|^2_{L^2(\Omega)} + \|\varphi\|^2_{L^2(\Omega)}. \tag{9.21}$$

From equation (9.14) we have

$$\|\varphi\|^2_{L_2(\Omega)} \leq \frac{1}{\mu_0} \|\Delta\varphi\|^2_{L^2(\Omega)}. \tag{9.22}$$

Using (9.22) in (9.21) we obtain

$$\frac{c_1}{1 + \frac{1}{\mu_0}} \|\varphi\|^2_{H^2_0(\Omega)} \leq \|\Delta\varphi\|^2_{L^2(\Omega)}.$$

∎

9.5   The Trace Theorem

For the applications of Sobolev spaces to boundary value problems it is necessary to restrict functions in Sobolev spaces to the boundary of the domain in order to satisfy the assigned boundary conditions. On the other hand, how smooth do our boundary data have to be so that functions in $H^m(\Omega)$ can take such values? In this section we will introduce the *trace operator* which restricts a function defined on the domain $\bar{\Omega}$ to the boundary of Ω, denoted Γ.

For instance, let $f \in C^\infty(\bar{\Omega})$. For such an f the trace is simply its restriction to Γ, which is well defined. Notice also that $C^\infty(\bar{\Omega})$ is a dense subspace of $H^k(\Omega)$. With this in mind, we have the theorem below.

THEOREM 9.11

Let k be any integer strictly greater than zero. The trace on Γ, defined at first on $C^\infty(\bar{\Omega})$, can be extended (in a unique manner) as a continuous linear map of $H^k(\Omega)$ onto $H^{k-\frac{1}{2}}(\Gamma)$.

PROOF   For a proof see Treves [29], p. 237.   ∎

Another version of Theorem 9.11 is given in Theorem 9.15 at the end of this section. But first we have to introduce the trace operator in R^n.

THEOREM 9.12

Let $s > \frac{1}{2}$, then there exists a continuous linear map

$$T : H^s(R^n) \to H^{s-\frac{1}{2}}(R^{n-1}),$$

called the trace operator, with the property that for any $\varphi \in C_0^\infty(R^n)$:

$$T(\varphi)(x_1, ..., x_{n-1}) = \varphi(x_1, x_2, ..., x_{n-1}, 0). \qquad (9.23)$$

PROOF　Let $\varphi \in C_0^\infty(R^n)$ and let $g(x') = \varphi(x', 0)$. Let $\tilde{\varphi}$ denote the Fourier transform of φ with respect to the mth variable only, i.e.,

$$\tilde{\varphi}(x', \xi_m) = \frac{1}{\sqrt{2\pi}} \int_{-\infty}^\infty \varphi(x', x_m) e^{-ix_m \xi_m} d\xi_m;$$

let $\hat{\varphi}$ and $\hat{g}$ denote the Fourier transform of φ and g in R^n and, respectively, R^{n-1}. The Fourier inversion formula yields

$$g(x') = \varphi(x', 0) = \frac{1}{\sqrt{2\pi}} \int_{-\infty}^\infty \tilde{\varphi}(x', \xi_m) d\xi_m.$$

Applying the Fourier transform, we find

$$\hat{g}(\xi') = \int_{-\infty}^\infty \hat{\varphi}(\xi) d\xi_m.$$

We now estimate

$$\|g\|_{H^{s-\frac{1}{2}}}^2 \leq C_1 \int_{R^{n-1}} |\hat{g}(\xi')|^2 (1 + |\xi'|^2)^{s-\frac{1}{2}} d\xi'$$

$$= \frac{C_1}{2\pi} \int_{R^{n-1}} \left| \int_{-\infty}^\infty \hat{\varphi}(\xi) d\xi_m \right|^2 (1 + |\xi'|^2)^{s-\frac{1}{2}} d\xi'$$

$$\leq C \int_{R^{n-1}} (1 + |\xi'|^2)^{s-\frac{1}{2}} \int_{-\infty}^\infty |\hat{\varphi}(\xi)|^2 (1 + |\xi|^2)^s d\xi_m$$

$$\times \int_{-\infty}^\infty (1 + |\xi|^2)^{-s} d\xi_m d\xi'. \qquad (9.24)$$

If $s > \frac{1}{2}$, we have

$$\int_{-\infty}^\infty (1 + |\xi|^2)^{-s} d\xi_m = \int_{-\infty}^\infty (1 + |\xi'|^2 + \xi_m^2)^{-s} d\xi_m$$

$$= (1 + |\xi'|^2)^{-s+\frac{1}{2}} \int_{-\infty}^\infty (1 + y^2)^{-s} dy.$$

By inserting into (9.24), we find that $\|T\varphi\|_{H^{s-\frac{1}{2}}} \leq C\|\varphi\|_{H^s}$ for every $\varphi \in C_0^\infty(R^n)$. Hence T can be extended by continuity to all of $H^s(R^n)$. The proof of Theorem 9.12 is now completed.　∎

Every function in $H^{s-\frac{1}{2}}(R^{n-1})$ can be obtained in this way by traces of functions in $H^s(R^n)$. This is stated in the theorem below.

THEOREM 9.13
Let $s > \frac{1}{2}$, then there exists a bounded linear mapping:

$$Z : H^{s-\frac{1}{2}}(R^{n-1}) \to H^s(R^n)$$

with the property that

$$TZ\varphi = \varphi \quad \text{for all } \varphi \in H^{s-\frac{1}{2}}(R^{n-1}).$$

The trace operator is surjective.

PROOF   For a proof see Wloka [31], p. 124.   ∎

By Theorem 9.13 we can construct a function $Z\varphi$ with a given trace φ. In the case where $s > \frac{1}{2} + k$ there exists a continuous trace operator T_k given by

$$T_k : H^s(R^n) \to \prod_{j=0}^{k} H^{s-j-\frac{1}{2}}(R^{n-1}). \tag{9.25}$$

For the trace operator in (9.25) we have a theorem analogous to Theorem 9.13 above. This is stated in the following theorem.

THEOREM 9.14
Let $s > \frac{1}{2} + k$, then there exists a bounded linear mapping:

$$Z_s : \prod_{j=0}^{k} H^{s-j-\frac{1}{2}}(R^{n-1}) \to H^s(R^n)$$

with the property such that

$$T_s Z_s \varphi = \varphi \text{ for all } \varphi \in C_0^\infty(R^{n-1}).$$

PROOF   A proof can be found in Renardy and Rogers [23], p. 221.
∎

Using the theorems above and a so-called "partition of unity argument", we can extend the results to domains with a bounded boundary.

THEOREM 9.15

Let $k > 0$ be an integer. If Ω is open, bounded, and of class C^∞, then there exists a bounded linear trace operator:

$$T : H^k(\Omega) \to H^{k-\frac{1}{2}}(\Gamma).$$

There also exists a bounded linear right inverse Z:

$$Z : C^\infty(\Gamma) \to C^\infty(\overline{\Omega})$$

so that

$$TZ = I \tag{9.26}$$

T is surjective.

PROOF   A proof can be found in Grubb [10], p. 3.24.   ∎

We have the following two theorems.

THEOREM 9.16

Let k,l be positive integers such that $k > l$. Let Ω be open, bounded, and of class C^k; then there exists a continuous trace operator

$$T_l : H^k(\Omega) \to \prod_{j=0}^{l} H^{k-j-\frac{1}{2}}(\Gamma)$$

with the property that for all $\varphi \in C^\infty(\Omega)$

$$T_l\varphi = (\varphi, \frac{\partial\varphi}{\partial\nu}, \dots, \frac{\partial^l\varphi}{\partial\nu^l}).$$

The kernel of the trace operator T_{l-1} is precisely $H_0^k(\Omega)$. This is stated in the theorem below.

THEOREM 9.17

Let Ω be of class C^k and let Γ be bounded; then H_0^k is the set of all those functions u in $H^k(\Omega)$ for which

$$T_k u = 0,$$

or alternatively,

$$u = \frac{\partial u}{\partial\nu} = \cdots = \frac{\partial^{k-1}u}{\partial\nu^{k-1}} = 0 \tag{9.27}$$

on Γ in the sense of trace.

PROOF

If $u \in C_0^\infty(\Omega)$, it is clear that (9.27) holds. By continuity, (9.27) then holds for $u \in H_0^k(\Omega)$. We need to establish the converse. By using a partition of unity and local coordinate transformations, we are reduced to the case $\Omega = R_+^n$. Let now $k = 1$ and let $u \in H^1(R^n)$ be such that $u(x', 0) = 0$ in the sense of trace. Let Eu be the extension of u by zero. To show that $Eu \in H^1(R^n)$, it suffices to establish that $\frac{\partial Eu}{\partial x_i} = E(\frac{\partial u}{\partial x_i})$. This is clear for $i < n$. For $i = n$ we have for any $\varphi \in C_0^\infty(R^n)$:

$$
\int_{R^n} \frac{\partial \varphi}{\partial x_n}(Eu)dx = \int_{R_+^n} \frac{\partial \varphi}{\partial x_n} u dx
$$
$$
= \int_{R^{n-1}} \varphi(x', 0)u(x', 0)dx' - \int_{R_+^n} \varphi \frac{\partial u}{\partial x_n} dx
$$
$$
= -\int_{R^n} \varphi E \frac{\partial u}{\partial x_n} dx.
$$

An analogous argument applies to higher derivatives. Once we know that $Eu \in H^k(R^n)$, the rest follows by considering the sequence $u_p = Eu(x - \frac{1}{p}e_n)$. Since the support of u_p is bounded away from ∂R_+^n, it is easy to approximate u_p by C_0^∞-functions. ∎

Notice that if we use the trace operator T, a *boundary operator* can be written as

$$
B_j(x, D) = T[\sum_{|s| \leq m_j} \alpha_{j,s}(x)D^s], \quad \text{where } a_{j,s} \in C^\infty(\overline{\Omega}).
$$

9.6   Negative Sobolev Spaces and Duality

Since $H^k(\Omega)$ are Hilbert spaces, they are isometric to their dual spaces according to Riesz Representation Theorem. Hence, every linear functional on $H^k(\Omega)$ can be represented by the form $l(u) = (u, v)_{H^k(\Omega)}$. However, since $C_0^\infty(\Omega)$ is not dense in $H^k(\Omega)$, the linear functionals are not necessarily distributions. We are therefore obliged to make the following definition.

DEFINITION 9.8   *By $H^{-k}(\Omega)$ we denote the set of all linear functionals on $H_0^k(\Omega)$. Moreover, if M is R^n or a compact manifold of class C^k, $k \geq s$, then $H^{-s}(M)$ denotes the dual space of $H^s(M)$.*

Since $C_0^\infty(\Omega)$ is dense in $H_0^k(\Omega)$, $H^{-k}(\Omega)$ is a set of distributions. By Riesz Representation Theorem we know that for $f \in H^{-k}(\Omega)$ there exists a unique $u \in H_0^k(\Omega)$ such that

$$(f, v) = (u, v)_{H^k(\Omega)} \tag{9.28}$$

for every $v \in H_0^k(\Omega)$. We find the following relationship between f and u. For any $\varphi \in C_0^\infty(\Omega)$

$$(f, \varphi) = \sum_{|\alpha| \le k} (D^\alpha u, D^\alpha \varphi) = \sum_{|\alpha| \le k} (-1)^{|\alpha|} (D^{2\alpha}, u), \tag{9.29}$$

i.e.,

$$f = \sum_{|\alpha| \le k} (-1)^{|\alpha|} D^{2\alpha} u. \tag{9.30}$$

For any given $f \in H^{-k}(\Omega)$ there exists, therefore, a unique $u \in H_0^k(\Omega)$ satisfying (9.30). By interpreting $H_0^k(\Omega)$ as a boundary condition on Γ (as seen in Theorem 9.12) we have a useful tool that can be applied in the theory of elliptic boundary value problems.

The following inclusion relations are useful to keep in mind.

$$C_0^\infty(\Omega) \subset H_0^{k'}(\Omega) \subset H_0^k(\Omega) \subset L^2(\Omega) \subset H^{-k}(\Omega) \subset H^{-k'}(\Omega) \subset D'(\Omega) \tag{9.31}$$

when $k' \ge k$, each space being dense in the following spaces (except $D'(\Omega)$). See Aubin [3], p. 187, for a proof of (9.31).

Note the following statements about differentiation of distributions in negative Sobolev spaces.

- Let $u \in H^k(\Omega)$, $k \in Z$. Then $\frac{\partial u}{\partial x_i} \in H^{k-1}(\Omega)$.

- Let $f \in H^{-k}(\Omega)$, $k \in N$. Then there exist functions $g_\alpha \in L^2(\Omega)$ such that

$$f = \sum_{|\alpha| \le k} D^\alpha g_\alpha.$$

Chapter 10

Interpolation Spaces

The theory of interpolation spaces is a branch of functional analysis which has been applied to a number of areas in analysis, most notably to the theory of partial differential equations. We have in the previous chapter studied H^s Sobolev spaces and traces in H^s since they are "basic tools" in the analysis of boundary value problems. This leads in a natural way to the theory of interpolation of linear operators and function spaces. This theory plays a significant role in obtaining regularity results for boundary value problems.

The aim of this chapter is to present those results of interpolation theory directly relevant to the study of boundary value problems. The theory of the spectral decomposition of operators plays an important role in the proofs of some of the theorems in this chapter. Since a comprehensive presentation of the theory of spectral decomposition is beyond the scope of this book, we refer the reader to Diximier [6], Riesz and Nagy [24], and Yosida [32] for this. The reader can also consult Bergh and Löfström [5] for a more general presentation of interpolation spaces. Since the purpose of this chapter is only to introduce some fundamental concepts and properties of interpolation spaces, many proofs are omitted, but we have listed some relevant references.

10.1 Intermediate and Interpolation Spaces

Let A and B be two normed vector spaces that are compatible, i.e., there exists a topological Hausdorff vector space $\mathcal{U}$ such that A and B are subspaces of $\mathcal{U}$. Then we can form their intersection $A \cap B$ and their sum $A + B$. Their sum consists of all $u \in \mathcal{U}$ such that we can write $u = a + b$ for some $a \in A$ and $b \in B$.

A normed vector space C will be called *an intermediate space* between

A and B if

$$(A \cap B) \subset C \subset (A + B) \tag{10.1}$$

with continuous inclusions.

The normed vector space space C is called an *interpolation space* between A and B if, in addition to (10.1), there exists a linear bounded operator T such that

$$\begin{cases} T : A \to A \\ \\ T : B \to B \end{cases} \quad \text{implies } T : C \to C. \tag{10.2}$$

10.1.1   The Operator L.

Let X and Y be Hilbert spaces. In order to simplify the presentation, and since it satisfies our needs, we assume them to be separable, as always, in this book. Let

$$X \subset Y, \quad \text{where we have a continuous dense injection.} \tag{10.3}$$

Let $(\cdot, \cdot)_X$ and $(\cdot, \cdot)_Y$ denote scalar products in X and Y, respectively.

The space X may be defined as the domain of an operator L, i.e.,

$$D(L) = X \subset Y. \tag{10.4}$$

We assume that L is self-adjoint, positive, and unbounded in Y. The norm in X is equivalent to the graph norm

$$\left(\|u\|_Y^2 + \|Lu\|_Y^2 \right)^{\frac{1}{2}}, \quad u \in D(L) = X. \tag{10.5}$$

We will now illustrate a procedure which is closely linked to the variational formulation of elliptic boundary value problems.

We denote by $D(S)$ the set of $u's$ such that the mapping

$$v \to (u, v)_X, v \in X \tag{10.6}$$

is continuous in the topology induced by Y. Then

$$(u, v)_X = (Su, v)_Y. \tag{10.7}$$

We can then verify that

- $D(S)$ is dense in Y.

- S is a self-adjoint operator.

- S is strictly positive, $(Sv, v)_Y = \|v\|_X^2 \geq c\|v\|_Y^2$, $c > 0$.

Using spectral decomposition of self-adjoint operators, the powers of $S^\theta, \theta \in R$ or $\theta \in C$ are well defined. (See Yosida [32], Chapter XI, for the underlying theory.) In particular, we shall set

$$L = S^{\frac{1}{2}}. \tag{10.8}$$

The operator L is self-adjoint and positive in Y. From equations (10.7) and (10.8) we get

$$(u, v)_X = (Lu, Lv)_Y, \forall u, v \in X. \tag{10.9}$$

DEFINITION 10.1 *Under hypothesis (10.3) and L defined as (10.8) we set*

$$[X, Y]_\theta = D(L^{1-\theta}) \tag{10.10}$$

with the norm on $[X, Y]_\theta$ equal to the norm of the graph of $L^{1-\theta}$

$$\left(\|u\|_Y^2 + \|L^{1-\theta}u\|_Y^2\right)^{\frac{1}{2}}. \tag{10.11}$$

From the properties of spectral resolution (cf. Yosida [32], Chapter XI), X is dense in $[X, Y]_\theta$.

10.2 Intermediate Derivatives Theorem

The Intermediate Derivatives Theorem (Theorem 10.2) is a theorem that is frequently used in applications. In this section we will state the theorem.

First we define some terms and concepts which are used in the proof of the Intermediate Derivatives Theorem.

DEFINITION 10.2 *Let a, $b \in R$ such that $a < b$ (finite or not), and let X denote a Hilbert space. By $L^p(a, b, X)$ we denote the space of functions where*

$$\int_a^b \|u(t)\|^p dt^{\frac{1}{p}} < \infty. \tag{10.12}$$

DEFINITION 10.3   *[W(a, b) spaces] Let a, b ∈ R such that a < b (finite or not). Let X and Y be Hilbert spaces that fulfill (10.3), and m an integer ≥ 1. We set*

$$W(a,b) = \left\{ u | u \in L^2(a,b,X), \frac{\partial^m u}{\partial t^m} = u^{(m)} \in L^2(a,b,Y) \right\}, \qquad (10.13)$$

the derivatives $u^{(m)}$ to be taken in the sense of distribution.

We can provide $W(a,b)$ with the norm

$$\|u\|_{W(a,b)} = \left(\|u\|_{L^2(a,b;X)} + \|u^{(m)}\|_{L^2(a,b,Y)} \right)^{\frac{1}{2}}. \qquad (10.14)$$

It is easily verified that $W(a,b)$ is a Hilbert space.

LEMMA 10.1
Assume a, b ∈ R such that a < b and that at least one of them is finite. There exists a continuous linear operator p such that

$$p : W(a,b) \to W(-\infty, +\infty) \qquad (10.15)$$

and

$$p(u) = u \text{ a.e. on }]a, b[. \qquad (10.16)$$

PROOF   A proof can be found in Lions and Magenes [19], Vol. 1. pp. 13-14. ∎

We can now state the theorem.

THEOREM 10.2 (Intermediate Derivatives Theorem)
Let X and Y be Hilbert spaces that fulfill such that

$$X \subset Y \text{ with a continuous dense injection.} \qquad (10.17)$$

Let $[X,Y]_\theta$ be defined by (10.10), and W be defined as in (10.13). Then

$$u^{(j)} \in L^2(a, b, [X,Y]_{\frac{j}{m}}), \quad 1 \le j \le m - 1. \qquad (10.18)$$

Furthermore, $u \to u^{(j)}$ is a continuous linear mapping of

$$W(a,b) \to L^2(a, b; [X,Y]_{\frac{j}{m}}). \qquad (10.19)$$

REMARK 10.1   In the case $j = 0$, we simply have $u \in L^2(a, b, X)$ since $[X, Y]_0$ is X. ∎

The proof can be found in Lions and Magenes [19], Vol. I, pp. 16-18.

10.2.1   An Example

The following example is relevant for the study of elliptic and hyperbolic boundary value problems. Let V and H be two Hilbert spaces with

$$V \subset H \quad \text{with a continuous dense injection.} \qquad (10.20)$$

We identify H with its antidual H' (we abuse the notation by adopting identical symbols for both the dual and the antidual in order to avoid meaningless distinctions). We have, therefore,

$$V \subset H \subset V'. \qquad (10.21)$$

Now taking A to be the previously mentioned operator S, $V = X$, and $H = Y$, we have

$$D(A) \subset V \subset H \subset V', V = D(A^{\frac{1}{2}}). \qquad (10.22)$$

The following proposition demonstrates a situation which is relevant in the proof of *regularity theorems* for partial differential equations.

PROPOSITION 10.3
With V and H defined as above and if

$$u \in L^2(0, \infty, V) \text{ and } u'' \in L^2(0, \infty, V'), \qquad (10.23)$$

then $u' \in L^2(0, \infty, H)$.

PROOF   The result is a simple consequence of the Intermediate Derivatives Theorem. Simply choose $X = V$, $Y = V'$, and $m = 2$. ∎

THEOREM 10.4 (The Interpolation Inequality)
For every $u \in X$

$$\|u\|_{[X,Y]_\theta} \leq C \|u\|_X^{1-\theta} \|u\|_Y^\theta, \qquad (10.24)$$

where $C > 0$ is a constant.

PROOF   We have the inequality

$$\|L^{1-\theta}u\|_Y \leq \|Lu\|_Y^{1-\theta}\|u\|_Y^{\theta}. \tag{10.25}$$

Hence

$$
\begin{aligned}
\|u\|_{[X,Y]_\theta}^2 &= \left(\|u\|_Y^2 + \|L^{1-\theta}u\|_Y^2\right) \\
&\leq \left(\|u\|_Y^2 + \|Lu\|_Y^{2(1-\theta)}\|u\|_Y^{2\theta}\right) \\
&= \|u\|_Y^{2\theta}\left(\|u\|_Y^{2(1-\theta)} + \|Lu\|_Y^{2(1-\theta)}\right).
\end{aligned} \tag{10.26}
$$

Recall now that $\|u\|_X$ is equivalent to the graph norm $\left(\|u\|_Y^2 + \|Lu\|_Y^2\right)^{\frac{1}{2}}$. We have, therefore,

$$c\left(\|u\|_Y^2 + \|Lu\|_Y^2\right) \leq \|u\|_X^2.$$

Hence

$$c(\|u\|_Y^2 + \|Lu\|_Y^2)^{1-2\theta} \leq \|u\|_X^{2(1-\theta)}. \tag{10.27}$$

Hence

$$c\left(\|u\|_Y^{2(1-\theta)} + \|Lu\|_Y^{2(1-\theta)}\right) \leq \|u\|_X^{2(1-\theta)}, \tag{10.28}$$

where we have used the fact that for $a, b \in R_+ \cup \{0\}$ and $0 \leq p \leq 1$,

$$(a + b)^p \leq a^p + b^p. \tag{10.29}$$

Applying (10.28) in (10.26), we have the Interpolation Inequality (10.24).

∎

10.2.2   A Continuity Property

In this section we state and prove a theorem which states some properties of the elements of $W(a,b)$ and their behavior under differentiation with respect to t. We define first the following normed space.

DEFINITION 10.4   *If E is a Banach space, we set*

$C_b(a, b; E) = $ *continuous and bounded functions of* $]a, b[\to E$.

The space $C_b(a, b; E)$ is provided with the following norm

$$\|\varphi\|_{C_b(a,b;E)} = \sup_{t\in[a,b]} \|\varphi\|_E. \tag{10.30}$$

THEOREM 10.5 (Continuous Derivatives Theorem)
With the notation above, for $u \in W(a,b)$ we have

$$u^{(j)} \in C_b(a,b;[X,Y]_{(j+\frac{1}{2})/m}), \quad 0 \leq j \leq m-1, \quad (10.31)$$

$u \to u^{(j)}$ *being a continuous and linear mapping of*

$$W(a,b) \to C_b(a,b;[X,Y]_{(j+\frac{1}{2})/m}).$$

Some remarks are in order:

REMARK 10.2 We have

$$[X,Y]_{\frac{j}{m}} \subset [X,Y]_{(j+\frac{1}{2})/m}. \quad (10.32)$$

Therefore, from Theorems 10.2 and 10.5 we deduce that if $u \in W(a,b)$, then $u^{(j)}$ is square integrable with values in $[X,Y]_{\frac{j}{m}}$ and is continuous with values in a *larger* space. To be more precise, the function which is known to belong to $L^2(a,b;[X,Y]_{\frac{j}{m}})$ (Theorem 10.2) is actually, after a modification on a set of measure zero, a continuous mapping of

$$[a,b] \to [X,Y]_{(j+\frac{1}{2})/m}. \quad (10.33)$$

∎

PROOF See Lions and Magenes [19] Vol. I, p. 20. ∎

10.3 Interpolation Theorem

In this section we state the so-called *main theorem of interpolation*. This theorem provides us with the properties that will enable us to characterize $[X,Y]_\theta$ as "interpolation spaces". We will first define some necessary new spaces, and state some theorems which are necessary in proving the main theorem.

DEFINITION 10.5 *We define for $s \in R_+$ the following Hilbert space*

$$W(-\infty,+\infty;s;X,Y) = \{u \mid \hat{u} \in L^2(R_\tau;X), \mid \tau \mid^s \hat{u} \in L^2(R_\tau;Y)\}, \quad (10.34)$$

provided with the norm

$$\|u\|_{W(-\infty,+\infty;s;X,Y)} = \left(\|\hat{u}\|^2_{L^2(R_t;X)} + \| \; | \; \tau \; |^s \; \hat{u}\|^2_{L^2(R_t;Y)} \right)^{\frac{1}{2}}. \qquad (10.35)$$

LEMMA 10.6
For every $u \in W(-\infty, +\infty; s; X, Y)$, we have

$$u^{(j)}(0) \in [X,Y]_{(j+\frac{1}{2})/s}, \;\; 0 \le j < s - \frac{1}{2}. \qquad (10.36)$$

Furthermore, the mapping

$$u \to \{u^j(0)\}_{0 \le j < s - \frac{1}{2}} \qquad (10.37)$$

of

$$W(-\infty, +\infty; s; X, Y) \to \prod_{0 \le j < s - \frac{1}{2}} [X,Y]_{(j+\frac{1}{2})/s} \qquad (10.38)$$

is surjective.

PROOF   A proof can be found in Lions and Magenes [19] Vol. I, pp. 24-26.   ∎

REMARK 10.3   Since the mapping (10.37) is surjective $[X,Y]_{(j+\frac{1}{2})/s}$, $0 \le j < s - \frac{1}{2}$ may be defined as the space described by $u^j(0)$ as u describes $W(-\infty, +\infty; s; X, Y)$.   ∎

Let $\{\mathcal{X}, \mathcal{Y}\}$ be a couple of Hilbert spaces having properties identical to the Hilbert space couple $\{X, Y\}$ under consideration. Let π be a continuous linear operator of X into $\mathcal{X}$, and of Y into $\mathcal{Y}$, i.e.,

$$\pi \in \mathcal{L}(X; \mathcal{X}) \cap \mathcal{L}(Y; \mathcal{Y}). \qquad (10.39)$$

THEOREM 10.7 (Main Theorem of Interpolation)
If π satisfies 10.39, then

$$\pi \in \mathcal{L}([X,Y]_\theta; [\mathcal{X}, \mathcal{Y}]_\theta), \qquad \forall \theta, 0 < \theta < 1. \qquad (10.40)$$

PROOF   Let θ be fixed. Set $s = \frac{1}{2\theta}$; then the space $[X,Y]_\theta$ is the space described by $u(0)$ as u describes $W(-\infty, +\infty; s; X, Y) = W$ (cf. Remark

10.3). Therefore, if $a \in [X, Y]_\theta$, there exists a $u \in W$ such that $u(0) = a$. Define v by

$$v(t) = \pi u(t) \text{ a.e.} \tag{10.41}$$

Since $\pi \in \mathcal{L}(X; \mathcal{X})$, (resp. $\mathcal{L}(Y; \mathcal{Y})$),

$$\hat{v} = \pi \hat{u} \text{ and } \hat{u} \in L^2(R_\tau; X) \text{ implies } u \in L^2(R_t; X),$$

we see that

$$v \in L^2(R_t; \mathcal{X}) \text{ and resp. } |\tau|^s \hat{v}(\tau) = \pi(|\tau|^s \hat{u}(\tau)) \in L^2(R_\tau; \mathcal{Y}). \tag{10.42}$$

Let

$$\|\pi\|_{\mathcal{L}(\mathcal{X};\mathcal{X})} = \alpha,$$
$$\|\pi\|_{\mathcal{L}(\mathcal{X};\mathcal{Y})} = \beta.$$

Hence,

$$\|v\|_{L^2(R_t,\mathcal{X})} \leq \alpha \|u\|_{L^2(R_t,X)},$$
$$\| \, |\tau|^s \, v\|_{L^2(R_t,\mathcal{Y})} \leq \beta \|u\|_{L^2(R_t,X)}.$$

Hence we can conclude that $v \in W(-\infty, +\infty; s; \mathcal{X}, \mathcal{Y}) = \mathcal{W}$ and

$$\|v\|_\mathcal{W} \leq \max\{\alpha, \beta\} \|v\|_\mathcal{W}. \tag{10.43}$$

But since $v \to v(0)$ is a continuous mapping of $\mathcal{W} \to [\mathcal{X}, \mathcal{Y}]_\theta$, we have $v(0) \in [\mathcal{X}, \mathcal{Y}]_\theta$. But $v(0) = \pi u(0) = \pi a$. Hence,

$$\|\pi a\|_{[\mathcal{X},\mathcal{Y}]_\theta} \leq \|v\|_\mathcal{W}. \tag{10.44}$$

This together with the inequality (10.43) yields

$$\|\pi a\|_{[\mathcal{X},\mathcal{Y}]_\theta} \leq c \max(\alpha, \beta) \inf \|u\|_\mathcal{W}, \quad u(0) = a. \tag{10.45}$$

From this we obtain (10.40) ∎

10.3.1 Reiteration Properties

Let θ_0 and θ_1 be fixed in $]0, 1[$ with

$$\theta_0 < \theta_1. \tag{10.46}$$

Then

$$D(L^{1-\theta_0}) \subset D(L^{1-\theta_1}). \tag{10.47}$$

Hence

$$[X,Y]_{\theta_0} \subset [X,Y]_{\theta_1}. \tag{10.48}$$

PROPOSITION 10.8
The space $[X,Y]_{\theta_0}$ is dense in $[X,Y]_{\theta_1}$.

PROOF   See Lions and Magenes [19], Vol. I, p. 28.   ∎

THEOREM 10.9
Let θ_0 and θ_1 be fixed in $]0,1[$. Then for all $\theta \in]0,1[$ we have

$$[[X,Y]_{\theta_0},[X,Y]_{\theta_1}]_\theta = [X,Y]_{(1-\theta)\theta_0+\theta\theta_1}. \tag{10.49}$$

PROOF   See Lions and Magenes [19], Vol. I, pp. 28-29.   ∎

REMARK 10.4   The property (10.49) is the reiteration property. For successive applications of the "operation" $\{X,Y\} \to [X,Y]_\theta$, for various values of θ we recover a space of the same type for different values of the interpolation parameter.   ∎

10.3.2   Duality

Since $X \subset [X,Y]_\theta \subset Y$, each space being dense in the following one, we have, by duality,

$$Y' \subset [X,Y]'_\theta \subset X', \tag{10.50}$$

each space being dense in the following ones. We have the following duality theorem.

THEOREM 10.10
For all $\theta \in]0,1[$

$$[X,Y]'_\theta = [Y',X']_{1-\theta}. \tag{10.51}$$

PROOF   See Lions and Magenes [19], Vol. I, p. 29.   ∎

10.4 Interpolation between $H^s, s \in R_+$ Spaces

We have already seen how the H^s Sobolev spaces are defined for $s \in R$. In this section we shall see how interpolation between two H^s spaces is possible.

First we define H^s Spaces, $s \in R_+$ as interpolation spaces.

DEFINITION 10.6   *By definition we set*

$$H^s = [H^m(\Omega), H^0(\Omega)]_\theta, \quad (1-\theta)m = s \ \ m \in Z, \ 0 < \theta < 1. \quad (10.52)$$

REMARK 10.5

- Up to an equivalence in norms, Definition 10.6 depends only on s and not on the choice of m (as long as $(1-\theta)m = s$).

- Definition 10.6 gives us the usual definition for a Sobolev space (up to an equivalence in norms).

∎

THEOREM 10.11
For all $s_1, s_2 > 0$, $s_2 < s_1$, $0 < \theta < 1$, we have that

$$[H^{s_1}(\Omega), H^{s_2}(\Omega)]_\theta = H^{(1-\theta)s_1 + \theta s_2}(\Omega). \quad (10.53)$$

Proof:

Let m be an integer such that $m \geq \max(s_1, s_2)$. Then

$$H^{s_i}(\Omega) = [H^m(\Omega), H^0(\Omega)]_{\theta_i}, \quad (1-\theta_i)m = s_i, i = 1, 2. \quad (10.54)$$

Hence,

$$[H^{s_1}(\Omega), H^{s_2}(\Omega)]_\theta = \left[[H^m(\Omega), H^0(\Omega)]_{\theta_1}, [H^m(\Omega), H^0(\Omega)]_{\theta_2}\right]_\theta$$
$$= [H^m(\Omega), H^0(\Omega)]_{(1-\theta)\theta_1 + \theta\theta_2}. \quad (10.55)$$

The last step is simply a consequence of Theorem 10.9. Now, since

$$(1 - (1-\theta)\theta_1 - \theta\theta_2)\, m = ((1-\theta)(1-\theta_1) + \theta(1-\theta_2))\, m$$
$$= (1-\theta)s_1 + \theta s_2,$$

(10.53) follows from Definition 10.6, and the proof is completed.

10.5 Interpolation with Hilbert Range

In this section we list some properties and theorems of particular interest when studying time-dependent evolution problems. As in the preceeding, we assume that X is a Hilbert or Banach space.

We can verify that the space $\tilde{X} = C^0([0,T];X)$ (i.e., the space of continuous functions of $[0,T] \to X$) is a Banach space with the norm

$$\sup_{t \in [0,T]} \|f(t)\|_X. \tag{10.56}$$

We have the following theorem for the interpolation between two spaces of continuous functions with Hilbert range.

THEOREM 10.12
Let X and Y denote Hilbert spaces with

$$X \subset Y \quad \text{with a continuous dense injection.} \tag{10.57}$$

Then

$$[C([0,T];X), C([0,T];Y)]_\theta = C([0,T];[X,Y]_\theta). \tag{10.58}$$

Proof:
See Lions and Magenes [19] pp. 95-96.

10.5.1 The Space $L^p(0,T;X)$

Let $[0,T]$ be a bounded interval in $R_+ \cup \{0\}$ and let the values of p be in the interval $[1 \leq p \leq \infty]$. Let X be a Banach space. Then $f \in L^p(0,T;X)$ is the space of functions on $[0,T]$ with range X such that

$$\|f\|_{L^p(0,T;X)} = \left(\int_0^T \|f(t)\|_X^p dt \right)^{\frac{1}{p}} < \infty \quad \text{for } 1 \leq p < \infty, \tag{10.59}$$

and for $p = \infty$

$$\sup_{0 \leq T} \|f(t)\|_X < \infty. \tag{10.60}$$

Then $L^p(0,T;X)$ is the space of *p-norm integrable functions*, and it is a Banach space for the norms (10.59) and (10.60) As it was the case for the usual L^p-spaces, we can only talk of equivalence classes of functions, and the comments preceding Theorem 2.5 apply.

10.5.2 Distributions on $]0, T[$ with Values in X.

We define the space of distributions on $]0, T[$ with values in X, denoted $D'(]0, T[; X)$, as the space

$$D'(]0, T[; X) = B(C_0^\infty(]0, T[); X), \tag{10.61}$$

provided with the topology of uniform convergence on the bounded sets of $C_0^\infty(]0, T[)$.
From (10.61) it is important to note that, if $f \in D'(]0, T[; X)$, then for all $\varphi \in C_0^\infty(]0, T[)$, $\langle f, \varphi \rangle$ has its range in X and is a continuous mapping of $C_0^\infty(]0, T[)$ into X.

Example 10.1
If $f \in L^2(0, T; X)$, we define

$$\tilde{f} \in D'(]0, T[; X) \tag{10.62}$$

by

$$\langle \tilde{f}, \varphi \rangle = \int_0^T f(t)\varphi(t)dt \quad , \forall \varphi \in C_0^\infty(]0, T[). \tag{10.63}$$

Hence, we have a continuous linear mapping $f \to \tilde{f}$ of $L^2(0, T; X) \to D'(]0, T[; X)$. From this mapping we identify $\tilde{f}$ with f and obtain the result that

$$L^2(0, T; X) \subset D'(]0, T[; X). \tag{10.64}$$

Recall here that $\langle \tilde{f}, \varphi \rangle \in X$. □

10.5.3 Interpolation between $L^p(0, T; H)$ Spaces

We have the following theorem for the interpolation between two $L^p(0, T; H)$ spaces where H denotes a Hilbert space.

THEOREM 10.13
Let X and Y denote Hilbert spaces with

$$X \subset Y \quad \text{with a continuous dense injection.} \tag{10.65}$$

Then

$$[L^p(0, T; X), L^p(0, T; Y)]_\theta = L^p(0, T; [X, Y]_\theta). \tag{10.66}$$

Chapter 11

Linear Elliptic Operators and Variational Theory of Boundary Value Problems

In this chapter we present a short introduction to the theory of elliptic operators and related boundary value problems.

We also present the basic tools for treating boundary value problems on so-called variational form. We consider the *elliptic problem*

$$Au + \lambda u = f, \tag{11.1}$$

$$B_j u = 0 \ , \ 0 \leq j \leq m - 1. \tag{11.2}$$

To put Problem (11.1)-(11.2) above into the variational formulation means that we seek a unique solution $u \in V$ of

$$a(u, v) + \lambda(u, v) = (f, v) \quad , \forall v \in V, \tag{11.3}$$

where $a(u, v)$ is a sesquilinear form associated to the partial differential equation, and V is a prescribed space. We demand of the variational formulation not only that $a(u, v) = (Au, v)$, but also that u satisfies the boundary conditions (11.2) when $u \in V$. This implies that not all problems can be solved by a variational formulation.

In the first section of this chapter we present some classical definitions and results, while in the following sections we discuss the demands mentioned above in more detail.

We adopt in this and the following chapters the notation V' for the dual space V^*, which is more common in variational theory.

11.1   Elliptic Operators

Consider the following linear differential operator on the space R^n

$$A(D)u = \sum_{|p| \le l} a_p D^p u \qquad (11.4)$$

of order l with constant coefficients a_p. To this operator we associate a polynomial (*the characteristic form of* **A**) in $\xi = (\xi_1, ..., \xi_n) \in R^n$

$$A_0(\xi) = \sum_{|p|=l} a_p \xi^p, \qquad (11.5)$$

where $\xi^p = \xi_1^{p_1} \xi_2^{p_2} .... \xi_n^{p_n}$.

DEFINITION 11.1   *[Elliptic Operator] The above mentioned operator A is said to be elliptic if*

$$A_0 \ne 0 \; \forall \xi \in R^n \setminus \{0\}. \qquad (11.6)$$

PROPOSITION 11.1
For $n > 2$, every elliptic operator is of even order.

PROOF   See Lions and Magenes [19], Vol. II, pp. 109-110.   ∎

Now let $\Omega \subset R^n$ be open, and assume that the coefficients a_p are now functions defined on Ω ($\bar{\Omega}$). The characteristic form (11.5), which will also depend on x, is denoted by $A_0(x, \xi)$. The operator A is *elliptic* in Ω ($\bar{\Omega}$) if the operator $A(x, D)$, considered as an operator with constant coefficients $a_p(x)$, is elliptic for every $x \in \Omega$ ($\bar{\Omega}$).

Example 11.1

- Proposition 11.1 no longer holds if $n = 2$. The *Cauchy Riemann* operator $\frac{\partial}{\partial x_1} + i \frac{\partial}{\partial x_2}$ is elliptic in R^n and is of order 1. However, if the coefficients are real, the proposition holds for n = 2.

- Operators of the form Δ^{2q} and Δ^{2q+1} of orders $4q$ and $4q+2$, respectively, are elliptic. $q = 0$ in the latter gives the *Laplacian* (Δ).

□

REMARK 11.1   In view of Proposition 11.1, we shall from now on assume that the order of A is even ($= 2$m).   ∎

DEFINITION 11.2   *[Uniformly elliptic] We say that the elliptic operator $A(x, D)$ is uniformly elliptic in Ω if there exists a constant $c > 0$ such that*

$$\left| \sum_{|\alpha|=2m} a_\alpha(x)\xi^\alpha \right| \geq c \mid \xi \mid^{2m} \text{ for } x \in \Omega, \ \xi \in R^n. \tag{11.7}$$

DEFINITION 11.3   *[Divergence form] We say that an operator is written on divergence form if there are functions $a_{pq} : \Omega \to C$ such that*

$$Au = A(x, D)u = \sum_{|p|,|q| \leq m} D^p(a_{pq}(x)D^q u). \tag{11.8}$$

In order to write a general partial differential operator on divergence form, it is required that the coefficients possess a certain amount of smoothness. For some necessary conditions, when $a_{pq} \in R$, see Renardy and Rogers [23], (Lemma 8.7, p. 285).

REMARK 11.2   We will assume that all the operators considered in the following can be written divergence form. Notice, howeve,r that the divergence form of an operator is not unique.   ∎

11.2   The Dirichlet Problem, Types of Solutions

The Dirichlet problem is probably the most important elliptic boundary value problem, and a number of classical examples originate from the study of physics. We will in this section study some solutions of this problem, also from a more general viewpoint, but we begin by stating the classical Dirichlet problem.

DEFINITION 11.4   *We define the space $C_b^k(\Omega)$ as the subspace of $C^k(\Omega)$ consisting of functions with bounded derivatives of order $\leq k$.*

DEFINITION 11.5   *[A classical solution] Let $\Omega \subset R^n$ be a bounded* *domain and suppose $f \in C_b(\Omega)$ is given. A function*

$$u \in C_b^{2m}(\Omega) \cap C_b^{m-1}(\overline{\Omega})$$

is a classical solution *of the Dirichlet problem if*

$$A(x, D)u = \sum_{|p|, |q| \leq m} D^p(a_{pq}(x)D^q u) = f \text{ in } \Omega, \qquad (11.9)$$

and

$$D^\alpha u = 0 \text{ for } |\alpha| \leq m - 1 \text{ on } \Gamma. \qquad (11.10)$$

The idea now is to look at larger function spaces in order to guarantee the existence of solutions to problems in such spaces. This is also the case with the classical Dirichlet problem. In order to use some of the standard solution methods, we must restate the problem in terms of Sobolev spaces.

DEFINITION 11.6   *[A strong solution] Let $\Omega \subset R^n$ be a bounded domain and suppose $f \in L^2(\Omega)$ is given. A function*

$$u \in H^{2m}(\Omega) \cap H_0^m(\Omega)$$

is a strong solution *of the Dirichlet problem if*

$$A(x, D)u = \sum_{|p|, |q| \leq m} D^p(a_{pq}(x)D^q u) = f \text{ in } \Omega. \qquad (11.11)$$

Note the following:

- For classical solutions, the differential equation (11.9) holds pointwise. For strong solutions, the differential equations are taken to hold in terms of equivalence classes (the right and left side of the equation represent the same equivalence class).

- Instead of explicitly imposing homogeneous boundary conditions, we include them implicitly by demanding that the solutions lie in H_0^m spaces.

- A classical solution to the Dirichlet problem is also a strong solution. This is simply a consequence of the preceding two observations.

We will now weaken the conditions further. In order to do so, we define a sesquilinear form by using integration by parts. Let $v \in H_0^m(\Omega)$ and $u \in H^{2m}(\Omega)$, then

$$\int_\Omega A u \bar{v} dx = \sum_{|p|, |q| \leq m} \int_\Omega a_{pq} D^p u \overline{D^q v} dx. \qquad (11.12)$$

Inspired by this, we can now define the following sesquilinear form.

DEFINITION 11.7   *We define*

$$a(u,v) = \int_\Omega \sum_{|p|,|q|\leq m} a_{pq}(x) D^p u \overline{D^q v} dx \qquad (11.13)$$

to be the sesquilinear form *associated with the divergence form of the elliptic partial differential operator A.*

Notice that the sesquilinear form $a(u,v)$ is defined for u and $v \in H^m(\Omega)$. Motivated by this, we define another type of solution of the Dirichlet problem.

DEFINITION 11.8   *[A weak solution] Let $\Omega \subset R^n$ be a bounded domain and suppose $f \in L^2(\Omega)$ is given. A function*

$$u \in H_0^m(\Omega)$$

is a weak solution *of the Dirichlet problem if*

$$a(u,v) = (f,v) \qquad (11.14)$$

for every $v \in H_0^m(\Omega)$.

Note that any strong solution to the Dirichlet problem is automatically a weak solution. Since we require much less smoothness of weak solutions than of strong ones, it is easier to prove the existence of such.

11.3 Boundary Operators

Consider $\Omega \subset R^n$, an open bounded set. We will now to introduce a theory for the study of boundary value problems for the operator A in H^m spaces. It will be necessary to impose some further conditions on both Ω and A. $\bar\Omega$ is now assumed to be compact with a boundary of class C^∞. A is assumed to be elliptic, of order $2m$, and written on the divergence form (11.8). Moreover, the coefficients a_{pq} are assumed to belong to class $C^\infty(\bar\Omega)$. A is therefore given by:

$$Au = A(x,D)u = \sum_{|p|,|q|\leq m} D^p(a_{pq}(x)D^q u), \qquad (11.15)$$

with

$$a_{pq} \in C^{\infty}(\bar{\Omega}). \tag{11.16}$$

11.3.1   Elliptic Equations

Consider the equations:

$$\begin{cases} Au = f & \text{in } \Omega, \\ B_j u = g_j & \text{on } \Gamma, \end{cases} \tag{11.17}$$

where the B_j's are a certain finite number of differential "boundary" operators and f and g_j are given.

Even for the simple cases it is not possible to arbitrarily assign the operators B_j's such that the problem is solvable. There is a need to formulate hypotheses which will classify those boundary problems, which will lead to *well-posed* problems. By a problem being well-posed (*in the sense of Hadamard*), we mean:

- There exists a solution.

- The solution is unique.

- The solution depends continuously on the data.

The last condition is a cause of some ambiguity. The meaning of the statement will have to be specified by the choice of norms in the context of each problem considered. In doing this, the hypotheses will neither be able to express independence of the boundary conditions from each other nor from the operator A.

Let B_j, $0 \le j \le v - 1$ be the v "boundary" operators defined by

$$B_j = B_j(x, D) = \sum_{|h| \le m_j} b_{jh}(x) D^h \tag{11.18}$$

with $b_{jh} \in C^{\infty}(\Gamma)$ and m_j being the order of B_j. Actually, one can consider $B_j \varphi$ as the operator

$$B_j(x, D)\varphi = \sum_{|h| \le m_j} b_{jh}(x) T\left[(D^h \varphi)\right], \tag{11.19}$$

φ being a function defined on $\bar{\Omega}$ for which $T(D^h \varphi)$ is the trace of $D^h \varphi$ on Γ.

DEFINITION 11.9   *[Normal system] The system of boundary operators* $\{B_j\}_{j=0}^{v-1}$ *is a normal system on* Γ *if :*

1. $m_j \neq m_i$ for $j \neq i$,

2. $m_j \leq 2m - 1$,

3. the leading order term in B_j contains a purely normal derivative, i.e., $B_j^p(x, \nu) \neq 0 \; \forall x \in \Gamma$, where ν is the unit outer normal.

Note that for $\upsilon = m$ the orders of the B_j's cover only half of the values from 0 to $(2m - 1)$. The missing orders can be filled by adding additional boundary operators $S_j, j = 1, ..., m$. In doing so, the conditions of normality still have to be fulfilled by the extended set of boundary operators.

We make the following hypotheses on the boundary operators:

1. there are m operators B_j,

2. the coefficients of B_j are infinitely differentiable on Γ,

3. $\{B_j\}_{j=0}^{m-1}$ is a normal system on Γ,

4. the order m_j of B_j is $\leq 2m - 1$.

Consider the system $\{F_i\}_{i=0}^{\upsilon-1}$ of υ differential boundary operators defined by

$$F_i \varphi = \sum_{|h| \leq m_i} f_{ih}(x) D^h \varphi, \qquad (11.20)$$

with $f_{ih} \in C^\infty(\Gamma)$.

DEFINITION 11.10 *[Dirichlet system] The system $\{F_i\}_{i=0}^{\upsilon-1}$ is a Dirichlet system of order υ on $\Gamma_1 \subseteq \Gamma$ if it is normal on Γ_1 and if the orders m_i run through exactly the set $0, 1, ..., \upsilon - 1$ when i goes from 0 to $\upsilon - 1$.*

11.3.2 The Formal Adjoint of A

Consider the operator A given by (11.15) and (11.16). By A^*, the *formal adjoint* of A, we denote the operator defined by :

$$A^* u = \sum_{|p|, |q| \leq m} D^p \overline{(a_{qp}(x) D^q u)}. \qquad (11.21)$$

Actually, since the following equation holds

$$\int_\Omega A u \bar{v} dx \; - \; \int_\Omega u \overline{A^* v} dx \; = 0 \quad \forall u, v \in C_0^\infty(\Omega), \qquad (11.22)$$

A^* is the adjoint of A in the sense of distributions, considered as an unbounded operator in $L^2(\Omega)$ with domain $C_0^\infty(\Omega)$.

We can now state a fundamental theorem on which most variational formulations of boundary value problems rely.

THEOREM 11.2 (The Theorem on Green's Formula)
Let A be an operator defined by (11.15) and (11.16) and assume it to be elliptic; let $\{B_j\}_{j=0}^{m-1}$ be a normal system on Γ given by (11.18). It is possible to choose nonuniquely another system of boundary operators $\{S_j\}_{j=0}^{m-1}$ normal on Γ, the S_j's having infinitely differentiable coefficients on Γ and being of order $\mu_j \leq 2m - 1$, such that the system $\{B_0, ..., B_{m-1}, S_0, ..., S_{m-1}\}$ is a Dirichlet system of order $2m$ on Γ. Having made this choice, there exist $2m$ boundary operators C_j and T_j, $j = 0, ..., m-1$, uniquely defined having the properties:

1. *the coefficients of C_j and T_j are in $C^\infty(\Gamma)$,*

2. *the order of C_j is $2m-1-\mu_j$ and the order of T_j is $2m-1-m_j$,*

3. *C_j and T_j form a Dirichlet system of order $2m$ on Γ. Moreover, the following Green's formula holds:*

$$\int_\Omega Au\bar{v}dx \; - \int_\Omega uA^*\bar{v}dx = \sum_{j=0}^{m-1} \int_\Gamma S_ju\overline{C_jv}d\Gamma \; - \sum_{j=0}^{m-1} \int_\Gamma B_ju\overline{T_jv}d\Gamma$$

$$(11.23)$$

for all u and v belonging to $H^{2m}(\Omega)$.

PROOF   See Lions and Magenes [19], Vol. I pp. 114-120.   ■

11.3.3   A Modified Version of Green's Formula

Let A be an elliptic operator given by (11.15) and (11.16). We consider the sesquilinear form $a(u, v)$ given by (11.13).

THEOREM 11.3
Let A and $a(u, v)$ be given as above. Let $\{B_j\}_{j=0}^{m-1}$ be a Dirichlet system of order m. Given that Ω and the coefficients of the operators are sufficiently smooth, there exist normal boundary operators C_j of order $2m-1-ord(B_j)$ such that for all $u, v \in H^{2m}(\Omega)$ we have:

$$a(u, v) = \int_\Omega (Au)\bar{v}dx - \sum_{j=0}^{m-1} \int_\Gamma (C_ju)\overline{(B_jv)}d\Gamma.$$

$$(11.24)$$

The operators $\{C_j\}$ are called "complementary boundary operators".

PROOF   See Lions and Magenes [19], Vol. I p. 120.   ∎

Example 11.2

Let A be a symmetric elliptic operator of order $2m$. We have, therefore, $a_{pq} = \overline{a_{qp}}$ and $a(u,v) = \overline{a(v,u)}$ since $A = A^*$. Given u and v in $H^{2m}(\Omega)$, we have in the notation of Theorem 11.3

$$a(u,v) = \int_\Omega (Au)\bar{v}dx - \sum_{j=0}^{m-1} \int_\Gamma (C_j u)(\overline{B_j v})d\Gamma. \qquad (11.25)$$

$$\overline{a(v,u)} = \int_\Omega (\overline{Av})u dx - \sum_{j=0}^{m-1} \int_\Gamma (\overline{C_j v})(B_j u)d\Gamma. \qquad (11.26)$$

Subtracting equation (11.25) from equation (11.26) we get

$$0 = \overline{a(v,u)} - a(u,v)$$

$$= \int_\Omega \overline{Av}u dx - \int_\Omega Au\bar{v}dx + \int_\Gamma \sum_{j=0}^{m-1} (C_j u)(\overline{B_j v})d\Gamma - \int_\Gamma \sum_{j=0}^{m-1} (\overline{C_j v})(B_j u)d\Gamma.$$

Hence,

$$\int_\Omega \left(\overline{Av}u - (Au)\bar{v}\right) dx = \int_\Gamma \sum_{j=0}^{m-1} (\overline{C_j v})(B_j u)d\Gamma - \int_\Gamma \sum_{j=0}^{m-1} (C_j u)(\overline{B_j v})d\Gamma. \qquad (11.27)$$

∎

11.4   V-elliptic and V-coercive Forms

In this section we state the important *Lax-Milgram Theorem*, define V-elliptic and V-coercive operators, and use the Lax-Milgram Theorem to put the original elliptic problem into the variational formulation.

THEOREM 11.4
Lax-Milgram Theorem

Let H be a Hilbert space and $a(u,v) : H \times H \to C$ a sesquilinear map for which

$$|a(u,v)| \le c_1\|u\|\|v\|, \quad u,v \in H \text{ (continuity)}$$

and

$$|a(u, u)| \geq c_2\|u\|^2, \quad u \in H \text{ (strictly positive)} \tag{11.28}$$

where $c_1, c_2 > 0$. Then there exists a unique bijective linear map $B : H \to H$, continuous in both directions and uniquely determined by the sesquilinear form $a(u, v)$, with

$$a(u, v) = (Bu, v), \quad a(B^{-1}u, v) = (u, v), \quad u, v \in H,$$

and for the norms we have $\|B\| \leq c_1$, $\|B^{-1}\| \leq \frac{1}{c_2}$.

PROOF   See Wloka [31] p. 272.
∎

In the following let V and H be Hilbert spaces such that

$$V \hookrightarrow H \hookrightarrow V', \tag{11.29}$$

where $\hookrightarrow$ means a continuous dense injection. We will refer to (11.29) as the *Gelfand triple*. In the following, let $a(u, v)$ be a sesquilinear form on V.

DEFINITION 11.11   V-elliptic
We say that the sesquilinear form $a(u, v)$ is V-elliptic if the following conditions are satisfied:

- *$|a(u, v)| \leq c_1\|u\|_V\|v\|_V$, $\quad \forall u, v \in V$, (continuity)*

- *$|a(u, u)| \geq c_2\|u\|_V^2$, $\quad \forall u \in V$,*

where c_1 and c_2 are independent of u and v.

REMARK 11.3   Some authors use Re $a(u, v) \geq c_2\|u\|_V$, $\quad \forall u, v \in V$ instead of the second assertion in Definition 11.11, but the definition here is simpler and perfectly adequate for our needs since we assume that $a(u, v) = \overline{a(u, v)}$.   ∎

Thanks to the Gelfand triple we can let $(u', u)_H$ be the representation of the functionals from V' by means of the scalar product $(\cdot, \cdot)_H$. By Riesz' Representation Theorem, we know that there exists an isomorphism $R : V' \to V$ such that

$$(u', u)_H = (Ru', u)_V. \tag{11.30}$$

From (11.30) and the continuity of $a(u, v)$ we have

$$a(u, v) = (Lu, v)_V = (R^{-1}Lu, v)_H, \quad u, v \in V. \tag{11.31}$$

In other words, the continuity of the sesquilinear form $a(u,v)$ is equivalent to the existence and continuity of a *representation operator* $A = R^{-1}L$,

$$A : V \to V', \quad a(u,v) = (Au,v)_H, u,v \in V. \tag{11.32}$$

Then we interpret the weak equation of $Au = f, f \in V'$ as $a(u,v) = (f,v)_H, \forall v \in V$ and we are able to present a solution theorem.

THEOREM 11.5

If the form $a(u,v)$ is V-elliptic, then the problem

$$a(u,v) = (f,y)_H \quad , \forall v \in V$$

possesses for each $f \in V'$ a unique solution $u \in V$, and this solution depends continuously on f. In other words, $A : V \to V'$ is a linear topological isomorphism between the spaces V and V'.

PROOF We have $A = R^{-1}L$. By the Lax-Milgram Theorem $L : V \to V$ is an isomorphism, and by Riesz' Representation Theorem $R : V' \to V$ is an isomorphism. Hence $L : V \to V'$ is also an isomorphism. ∎

The variational formulation of the elliptic problem (11.1)-(11.2) is

$$a(u,v) + \lambda(u,v) = (f,v) \quad , \forall v \in V, \tag{11.33}$$

which gives sense to the definition of a V-coercive sesquilinear form $a(u,v)$ stated below.

DEFINITION 11.12 *We say that a continuous sesquilinear form $a(u,v)$ is V-coercive if the following condition is satisfied: There exist constants $\lambda_0 \in R$ and $c_2 \geq 0$ with $|a(u,u)| + \lambda_0\|u\|_H^2 \geq c_2\|u\|_V^2$ for all $u \in V$ (Gårding's inequality).*

From this we obtain a result similar to the one of Theorem 11.5.

THEOREM 11.6

If the form $a(u,v)$ is V-coercive, then the problem

$$a(u,v) + \lambda(u,v)_H = (f,v)_H \quad , \forall v \in V$$

admits a unique solution for all $\lambda \in C$ satisfying

$$\text{Re } \lambda \geq \lambda_0. \tag{11.34}$$

Theorem 11.6 states the existence of a unique solution of (11.33). This means that there exists a representation operator A_V such that $a(u,v) = (A_V u, v)$, and we can therefore replace A by A_V in Problem (11.1)-(11.2). In order to ensure that a solution of the variational formulation (11.33) is also the solution of Problem (11.1)-(11.2), we have to study the difference between the sesquilinear form $a(u,v)$ and (Au,v) given by the modified Green's formula (Theorem 11.3). Here again, the continuity of the sesquilinear form $a(u,v)$ is equivalent to the existence and continuity of a representation operator of the sesquilinear form as mentioned above.

In this section we have only treated time-independent sesquilinear forms $a(u,v)$, but the results here also hold true for time-dependent sesquilinear forms if the time t is fixed. If we add some additional assumptions which, "loosely speaking", means that the sesquilinear form $a(t;u,v)$ is well-behaved on $[0;T]$, for fixed $u,v \in V$ we can show that the representation operator A is linear and continuous as a map

$$A : L^2(]0,T[;V) \to L^2(]0,T[;V').$$ (11.35)

We have not explicitly mentioned the boundary conditions, but they will be included in the definition of the space V, as we shall see in Section 11.5.

Example 11.3

We will show that we have a continuous extension of $(\cdot,\cdot)_H$ on $V' \times V$ which can be regarded as a new representation for the functionals from V'. We consider the Gelfand Triple

$$V \overset{i}{\hookrightarrow} H \overset{i'}{\hookrightarrow} V',$$ (11.36)

where $\hookrightarrow$ means a continuous dense injection. The imbeddings i, i' are continuous, injective, and have dense images in H and V', respectively. Note also that i' is the dual map of i.

If $v \in V$ and $h \in H$, we get from the definition of i' that

$$\langle i'h, v \rangle_{V',V} = \langle h, iv \rangle_{V',V} = (h, iv)_H.$$ (11.37)

From this we have

$$|(h, iv)_H| = |\langle i'h, v \rangle| \le \|i'h\|_{V'} \|v\|_V \le \|h\|_H \|v\|_V.$$ (11.38)

We may therefore consider each element $h \in H$ as an antilinear, continuous functional on V. We also have that $R(i') = i'H$ is dense in V' with respect to the functional norm $\|\cdot\|_{V'}$, that is, we can uniformly approximate each functional $\langle v', \cdot \rangle_V$ on the unit ball of V by the scalar product $(h, i\cdot)_H$:

$$\langle v', v \rangle_V = \lim_{i'h \to v'} (h, i\cdot)_H = \lim_{i'h \to v'} \langle i'h, \cdot \rangle_V.$$ (11.39)

We conclude that we can regard the continuous extension of $(\cdot,\cdot)_H$ on $V' \times V$ as a new representation formula of the functionals from V'. □

11.5 The Boundary Conditions

Theorem 11.6 gives the existence of a unique solution of the variational formulation (11.33) of Problem (11.1)-(11.2). This means that there exists a representation operator A_V such that $a(u,v) = (A_V u, v)$. In order to ensure that a solution of the variational formulation (11.33) also is the solution of Problem (11.1)-(11.2), we have to study the difference between the sesquilinear forms $a(u,v)$ and (Au,v) such that, when the difference is zero, the boundary conditions are fulfilled. That is, we are searching for an A_V with a domain corresponding to the boundary conditions, and this leads us to look at *realizations* of the operator A.

11.5.1 Realization of the Operator A

Before we take a closer look at the realization of the operator A, let us recall that the operator A is assumed written on the divergence form, i.e.,

$$Au = \sum_{|p|,|q|\leq m} D^p(a_{pq}(x)D^q u). \tag{11.40}$$

We associate to A two particular closed, densely defined, and unbounded operators in $L^2(\Omega)$ called A_{min} and A_{max} in the following way:

$$A_{min} = \{\text{The closure of } A|_{C_0^\infty(\Omega)} \text{ as an operator in } L^2(\Omega)\}.$$

$$A_{max} = \{\text{The operator acting as } A \text{ in } D'(\Omega) \text{ with domain}$$
$$D(A_{max}) = \{u \in L^2(\Omega)|Au \in L^2(\Omega)\} \}.$$

From the definition of A_{min} and A_{max} it is obvious that $A_{min} \subset A_{max}$. An operator A_γ is called a *realization of* A if

$$A_{min} \subset A_\gamma \subset A_{max}. \tag{11.41}$$

Example 11.4
Let us show that $D(A_{min}) = H_0^2(\Omega)$ for $A = -\Delta$.

In Lemma 9.9 we have obtained for any $\varphi \in C_0^\infty(\Omega)$ that there exists positive constants c_1 and c_2 such that

$$c_1\|\varphi\|_{H^2(\Omega)}^2 \leq \|\Delta\varphi\|_{L^2(\Omega)} + \|\varphi\|_{L^2(\Omega)}^2 \leq c_2\|\varphi\|_{H^2(\Omega)}^2.$$

Since we take the closure of A in the graph norm, we conclude from the definition of A_{min} that $D(A_{min}) = H_0^2(\Omega)$. For $D(A_{max})$ we can show that $D(A_{max}) \subset H^2_{loc}(\Omega)$. Notice here that this gives no control over the functions close to the boundary. In general, we can say that $D(A_{max})$ does not have the same smoothness properties as $D(A_{min})$. □

In the example above we concluded that $D(A_{min}) = H^2_0(\Omega)$ in the case of $A = -\Delta$ without considering boundary conditions. What we wish to find is a version of the operator A such that we solve Problem (11.1)-(11.2). Lemma 11.7 below states that the operator A_V defined from the variational formulation is a realization of A. We shall show later that A_V with the domain defined in Lemma 11.7 is the correct version of A under some additional hypotheses.

LEMMA 11.7
Assume that the sesquilinear form $a(u, v)$ is V-coercive on a space V satisfying

$$H^m_0(\Omega) \subset V \subset H^m(\Omega).$$

Moreover, let the operator A_V be defined by the variational formulation. Then A_V is a realization of A and its domain satisfies

$$D(A_V) = \{u \in D(A_{max}) \cap V \mid (Au, v)_{L^2(\Omega)} = a(u, v),\ \forall v \in V\}.$$

REMARK 11.4 A_V acts like A, but A_V is not unique since the divergence form of an operator in not unique. ■

PROOF   When $u \in D(A_V)$, we have that $(A_V u, v)_{L^2(\Omega)} = a(u, v)$ for all $v \in V$. When this is used in particular for $v \in C^\infty_0(\Omega)(\subset H^m_0(\Omega) \subset V)$, we see that $A_V u$ equals the distribution Au since

$$\langle Au, \bar{v} \rangle = \sum_{|p|,|q| \leq m} \langle D^p a_{pq} D^q u, \bar{v} \rangle$$

$$= \sum_{|p|,|q| \leq m} \langle a_{pq} D^q u, \overline{D^p v} \rangle = a(u, v)\ \forall u \in D(A_{max}),\ v \in C^\infty_0(\Omega).$$

So it follows (since u and $A_V u \in L^2(\Omega)$) that $u \in D(A_{max})$. This shows the inclusion "$\subset$". The inclusion "$\supset$" follows from the definition of A_V. Since $C^\infty_0(\Omega)$ is contained in $D(A_V)$, and A_V is closed, $A_V \supset A_{min}$. ■

Example 11.5
If we take $A = -\Delta$ and use Lemma 11.7 with $V = H^1_0(\Omega)$, we conclude that

$$D(A_V) = \{u \in D(A_{max}) \cap H^1_0(\Omega)\}.$$

The property that $u \in H_0^1(\Omega)$ represents the Dirichlet condition. That is, we have formulated the Dirichlet problem:

$$\begin{cases} -\Delta u = f \in L^2(\Omega), \\ u \mid_\Gamma = 0. \end{cases}$$

Since $u \in H_0^1(\Omega)$, we have $u \mid_\Gamma = 0$. This means that $D(A_V)$ corresponds to the boundary conditions.   ☐

11.5.2   The V-space

To get a closer look at the desired domain of the realization A_V of A, and in particular the space V, it is necessary to consider the modified Greens formula which expresses the difference between $a(u,v)$ and $(Au,v)_H$.
The modified Green's formula contains the boundary conditions with order less than $m - 1$. So if we say that there exists p boundary operators $\{B_0,....,B_{p-1}\}$ with $\mathrm{ord}(B_j) \leq m - 1$ in Problem (11.1)-(11.2) and apply the modified Greens formula (Theorem 11.3), we obtain

$$a(u,v) = \int_\Omega (Au)\bar{v}dx - \sum_{j=p}^{m-1} \int_\Gamma (C_j u)\overline{(B_j' v)}d\Gamma, \qquad (11.42)$$

if $\{B_0 v = 0,....,B_{p-1} v = 0\}$. Recall here that B_j''s are normal boundary operators which always exist such that $\{B_1,....,B_{p-1}, B_p',....,B_{m-1}'\}$ forms a Dirichlet system.

Example 11.6
Consider the problem

$$\begin{cases} Au = f \text{ in } \Omega, \\ T_{m-1}u = 0 \text{ on } \Gamma, \end{cases} \qquad (11.43)$$

where

$$T_{m-1} = (I, \frac{\partial}{\partial \nu}, \cdots \frac{\partial^{m-1}}{\partial \nu^{m-1}}). \qquad (11.44)$$

If we put $V = \{B_0 v = 0,...,B_{m-1} v = 0\} = T_{m-1} v = 0$ (i.e. $V = H_0^m(\Omega)$), we obtain that (11.42) can be written as

$$a(u,v) = \int_\Omega (Au)\bar{v}dx,$$

and by the variational formulation we get

$$\int_\Omega (Au)\bar{v}dx = (f,v)_H \quad , \forall v \in V. \qquad (11.45)$$

Hence, $u \in V = H_0^m(\Omega)$ is the solution of the problem (11.43).

□

In Example 11.6 we introduced the space V as the one which consists of those v which satisfy the boundary conditions. This is, however, not the general procedure. In general, we define the space V as in the Definition 11.13 below.

DEFINITION 11.13   The V-space
Let $B_j u = 0$ for $0 \leq j \leq p - 1$ be the boundary conditions for Problem (11.1)-(11.2), where $\mathrm{ord}(B_j) \leq m - 1$. Then

$$V = \{v \in H^m \mid B_j u = 0 \text{ for } 0 \leq j \leq p - 1\}.$$

The reason for this definition will become evident after the next section. It must be regarded in connection with the fact that only the operators with order $\leq m - 1$ appear in the modified Green's formula. This is also the reason why the boundary operators are divided into two groups according to their orders.

DEFINITION 11.14   *The boundary conditions are divided into two groups:*

- *The **stable conditions** for which $\mathrm{ord}(B_j) < m$.*

- *The **natural conditions** for which $\mathrm{ord}(B_j) \geq m$.*

11.6   The Variational Formulation of the Problems

In order to get the correct version of the realization A_V of A such that the variational formulation solves Problem (11.1)-(11.2), we have to put some restrictions on the orders of the boundary operators which appear in the problem. We can perceive these restrictions from the modified Green's formula as we now shall see.

Let us again consider Problem (11.1)-(11.2). It follows that, if u satisfies

$$a(u, v) = \int_\Omega f \bar{v} dx \quad \forall v \in V, \tag{11.46}$$

then

$$Au = f \quad \text{in } \Omega \tag{11.47}$$

and

$$\sum_{j=p}^{m-1} \int_{\Gamma} (C_j u)\overline{(B'_j v)} d\Gamma = 0 , \qquad \forall v \in V. \tag{11.48}$$

Since v is arbitrary and since $\{F_j\} = \{B_0,....,B_{p-1}, B'_p, ..., B'_{m-1}\}$ form a Dirichlet system of order m on Γ, it follows (see Lions & Magenes [19], Vol. I Lemma 2.2 p. 117) that

$$C_j u = 0 , \quad j = p,...,m-1. \tag{11.49}$$

Conversely, if u satisfies (11.47) and (11.49), then it follows from (11.42) that u also satisfies (11.46).

Thus we see how we can formally put the problem into a variational formulation, namely by assuming that we can choose $\{B'_p, ..., B'_{m-1}\}$ so that

$$C_j = B_j, \quad j = p,....,m-1 \tag{11.50}$$

and thereby getting rid of the boundary integral in (11.42) by using all of the boundary conditions. This means that the conditions

$$B_j u = 0, \quad j = p,....,m-1$$

also are fulfilled. But we cannot always do so, since this imposes restrictions. In the modified Green's formula (11.42), $\text{ord}(C_j) = 2m - 1 - \text{ord}(F_j)$. Therefore, if $\mu_p,....,\mu_{m-1}$ are numbers between 0 and $m-1$ so that $m_1,....,m_{p-1},\mu_p,....,\mu_{m-1}$ yield all numbers $0, 1,....,m - 1$ (in arbitrary order), then μ_j is the order of B'_j and thus the order of B_j, with $j = p,....,m-1$ given by $2m - 1 - \mu_j$. Hence, we have found the *restrictions on the order of the natural conditions* to be

$$m_j = 2m - 1 - \mu_j, \quad j = p,....,m-1, \tag{11.51}$$

and thereby also the conditions which ensure that the domain of A_V corresponds to the boundary conditions.

The "condition" (11.51) is a "restriction" since it is not necessarily satisfied for a regular elliptic problem as seen in Example 11.7 below.

Example 11.7
Let

$$Au = \Delta^2 u + u = \sum_{i,j=1}^{n} \frac{\partial^2}{\partial x_i^2} \left(\frac{\partial^2 u}{\partial x_j^2} \right) + u \tag{11.52}$$

with the boundary conditions given the boundary operators

$$B_0 u = u, \quad B_1 u = \frac{\partial \Delta u}{\partial \nu}. \qquad (11.53)$$

Then, we have $m = 2, p = 1, m_0 = 0, \mu_1 = 3$, and therefore (11.51) is not satisfied for $j = 1$. It is a regular elliptic problem, but it does not fit into the variational formulation. □

11.7 Assumptions for the Variational Problem

We will now extend the considerations to also include time-dependent problems. In order to do so, let us clarify our assumptions for the following. We consider Hilbert spaces V and H such that we have a *Gelfand triple*. We also demand the t-dependent sesquilinear form $a(t; u, v)$ to be *continuous*, in the sense that

- $|a(t; u, v)| \leq c\|u\|_V \|v\|_V$, $u, v \in V$ where c is independent of t.

We know from the preceding section that there exists a representation operator $A_V(t)$

$$A_V(t) : V \to V', \qquad (11.54)$$

which for each fixed t is continuous and linear, with $a(t; u, v) = (A(t)_V u, v)_H$. We assume, further, that

- $a(t; u, v)$ for $u, v \in V$ fixed is *continuously differentiable* with respect to t for $t \in [0, T]$ (T finite) and

$$\frac{d|a(t; u, v)|}{dt} \leq c\|u\|_V \|v\|_V, \quad \forall t \in [0, T], \qquad (11.55)$$

    c once again independent of t.

Then $a(t; u, v) \in C^1([0, T])$ for all $u, v \in V$, which implies that $a(t; u, v)$ is sufficiently well-behaved in order to have the situation for (11.35). We also assume that the sesquilinear form $a(t; u, v)$ is *V-coercive*. That is,

- There exist constants $\lambda_0, \alpha > 0$ such that

$$|a(t; u, u)| + \lambda_0 \|u\|_H^2 \geq \alpha \|u\|_V^2 \text{ for all } t \in [0, T], \text{ for all } u \in V. \qquad (11.56)$$

Finally, we assume that

- $a(t, u, v) = \overline{a(t, u, v)}$ for all $u, v \in V$.

We will consider the following problem:
Given $f \in L^2(0, T; H)$ (T finite) and initial conditions

$$u^0 \in V, \quad u^1 \in H,$$

we wish to find a function $u(t) \in L^2(0, T; V), u' \in L^2(0, T; H)$ such that in V' we have

$$\begin{cases} u'' + A(t)u = f, & t \in [0, T], \\ u(0) = u^0, \ u'(0) = u^1, \end{cases} \tag{11.57}$$

i.e.,

$$(u'', v)_H + (A(t)u, v)_H = (f, v)_H \quad \text{for all } v \in V. \tag{11.58}$$

(We use the notation u' for the t-derivative of u.)

With the assumptions in this section, this is the problem we will study in following sections.

11.8   A Classical Regularity Result

The classical result we present in this section will be highly relevant in obtaining the newer regularity results of second order hyperbolic systems. These newer results will be proved and commented upon in Section 12.2, Chapter 12.

The classical theorem, Theorem 11.8, is presented below in a more general setting than needed later. In Theorem 11.8 the boundary conditions are contained in the space V. This means that the space V below implies certain boundary conditions on the solution, and not necessarily that the representation operator may be identified with the operator A of the problem. For a discussion of this we refer to the previous sections.

11.8.1   A Regularity Theorem

THEOREM 11.8
Subject to the assumptions in Section 11.7, the problem

$$\begin{cases} u'' + A(t)u = f \text{ in } \Omega \times]0, T[, \\ u(x, 0) = u^0, u'(x, 0) = u^1 \text{ in } \Omega, \end{cases} \tag{11.59}$$

where

$$\begin{cases} F \in L^2(0, T; H), \\ u^0 \in V, \ u^1 \in H, \end{cases} \tag{11.60}$$

has a unique solution u. The map

$$\{F, u_0, u_1\} \to \left\{ u, \frac{\partial u}{\partial t} \right\} \tag{11.61}$$

is continuous and linear from

$$L^2(0, T; H) \times V \times H \to L^2(0, T; V) \times L^2(0, T; H). \tag{11.62}$$

REMARK 11.5   The result in Theorem 11.8 is also true if $F \in L^1(0, T; H)$, which is easily seen from the proof.   ∎

REMARK 11.6   As we shall see, we can extend the result (11.62) to a continuous linear mapping

$$L^2(0, T; H) \times V \times H \to L^\infty(0, T; V) \times L^\infty(0, T; H). \tag{11.63}$$

This result is also true if we replace $f \in L^2(0, T; H)$ with $f \in L^1(0, T; H)$.
∎

PROOF   We carry out the proof in three steps.

1. We approximate the solution by a sequence $u_m(t)$.

2. We show that $\|u_m\|_H^2 + \|u_m'\|_V^2 \le K$ (K a constant) and we can therefore extract a weakly convergent subsequence u_μ of u_m converging to $z(t)$, say.

3. We show that $z(t)$ is a solution of the problem.

For the proof of the uniqueness of the solution and the continuous linear mapping, we refer to Wloka [31], p. 437, for a comprehensive proof. We will only show the existence of a solution.

STEP 1 *Approximation by a sequence*
Since the spaces V and H are separable, we can let $\{w_n : n \in N\}$ be linearly independent and total in V, e.g., an orthonormal basis. Since V is dense in H, we have that H is the closure of the linear combinations of the w_n's in H. Then define

$$u_m^0 = \sum_{i=1}^m \xi_{im}^0 w_i, \quad u_{1m} = \sum_{i=1}^m \xi_{im}^1 w_i, \tag{11.64}$$

so that $u_{0m} \to u^0$ in V as $m \to \infty$ and $u_{1m} \to u^1$ in H as $m \to \infty$. Then it is natural to try to approximate the solution of the problem by a sequence $u_m(t)$ defined by

$$u_m(t) = \sum_{i=1}^{m} g_{im}(t)w_i,$$

where $g_{im}(t)$ is uniquely determined by the system of m linear ordinary differential equations stated below.

$$\frac{d^2}{dt^2}(u_m(t), w_j)_H + a(t; u_m(t), w_j) = (f(t), w_j)_H \quad 1 \le j \le m, \quad (11.65)$$

$$u_m(0) = u_{0m}, \quad u'_m(0) = u_{1m}. \quad (11.66)$$

From this we conclude that $u_m \in C^1([0, T]; V)$.

STEP 2 *An energy inequality*

We show now the energy inequality (11.67) below.

$$\|u'_m(t)\|_H^2 + \|u_m(t)\|_V^2 \le C, \quad (11.67)$$

where C is a constant to be estimated.

If we multiply (11.65) by $g'_{jm}(t)$ and take the sum over j, we obtain

$$(u''_m(t), u'_m(t))_H + a(t; u_m(t), u'_m(t)) = (f(t), u'_m(t))_H, \quad (11.68)$$

and hence

$$\frac{d}{dt}\left[\|u'_m(t)\|_H^2 + a(t; u_m(t), u_m(t))\right] = 2\mathrm{Re}(f(t), u'(t))_H. \quad (11.69)$$

By integration we obtain

$$\|u'_m(t)\|_H^2 + a(t; u_m(t), u_m(t)) = \|u_{1m}\|_H^2 + a(0; u_{0m}, u_{om})$$
$$+ \int_0^t a'(s; u_m(s), u_m(s))ds$$
$$+2\mathrm{Re}\int_0^t (f(s), u'_m(s))ds.$$

Now, since $a(u, v)$ is V-coercive and $a'(t; u, v)$ is continuous for all t, we obtain

$$\|u'_m(t)\|_H^2 + \alpha\|u_m(t)\|_V^2 \le k_0\|u_m(t)\|_H^2 + \|u_{1m}\|_H^2 + c\|u_{0m}\|_V^2$$
$$+c\int_0^t \|u_m(s)\|_V^2 ds + 2\int_0^t \|f(s)\|_H\|u'(s)\|_H ds,$$

i.e.,

$$\|u'_m(t)\|_H^2 + \|u_m(t)\|_V^2 \le c_1 \left\{ \|u_{0m}\|_V^2 + \|u_{1m}\|_H^2 + \|u_m(t)\|_H^2 \right\}$$
$$+ c_1 \int_0^t \|u_m(s)\|_V^2 + \|f(s)\|_H \|u'(s)\|_H ds.$$

From

$$\|u_m(t)\|_H \le \|u_{0m}\|_H + \int_0^t \|u'_m(s)\|_H ds \qquad (11.70)$$

it follows that

$$\|u_m(t)\|_H^2 \le 2\|u_{0m}\|_H^2 + 2 \left(\int_0^t \|u'_m(s)\|_H ds \right)^2 \le 2\|u_{0m}\|_H^2 + 2T \int_0^t \|u'_m(s)\|_H^2$$

If we define

$$W_m(t) = \|u'_m(t)\|_H^2 + \|u_m(t)\|_V^2,$$

it follows that

$$W_m(t) \le c_1 \left(\|u_{0m}\|_V^2 + \|u_{1m}\|_H^2 + 2 \left(\|u_{0m}\|_H^2 + T \int_0^t \|u'_m(s)\|_H^2 ds \right) \right)$$
$$+ c_1 \left(\int_0^t \|u_m(s)\|_V^2 ds + \int_0^t \|f(s)\|_H \|u'_m(s)\|_H ds \right). \qquad (11.71)$$

Since $2|ab| \le a^2 + b^2$ for $a, b \in R$, we have

$$W_m(t) \le c_2 \left(\|u_{0m}\|_V^2 + \|u_{1m}\|_H^2 + \int_0^t \|f(s)\|_H^2 ds \right) + c_2 \int_0^t W_m(s) ds.$$
$$(11.72)$$

If we now apply Gronwall's Lemma to (11.72) we have

$$\|u'(t)\|_H^2 + \|u_m(t)\|_V^2 \le c_3 \left(\|u_{0m}\|_V^2 + \|u_{1m}\|_H^2 + \int_0^T \|f(s)\|_H^2 ds \right).$$
$$(11.73)$$

STEP 3 *The solution of the problem*
In STEP 2 we obtained that

$$\|u'_m(t)\|_H^2 + \|u_m(t)\|_V^2 \le C \qquad (11.74)$$

with

$$C = c_3 \left(\|u_{0m}\|_V^2 + \|u_{1m}\|_H^2 + \int_0^T \|f(s)\|_H^2 ds \right). \qquad (11.75)$$

Hence, u_m is bounded in $L^2(0, T; V)$ and u'_m is bounded in $L^2(0, T; H)$. Then, since $L^2(0, T; V)$ and $L^2(0, T; H)$ are Hilbert spaces, we can extract subsequences $u_\mu(t)$ of $u_m(t)$ and $u'_\mu(t)$ of $u'_m(t)$ such that

$$u_\mu(t) \rightharpoonup z(t) \quad \text{weakly in } L^2(0, T; V) \text{ as } \mu \to \infty$$

and

$$u'_\mu(t) \rightharpoonup \tilde{z}(t) \quad \text{weakly in } L^2(0, T; H) \text{ as } \mu \to \infty.$$

Hence,

$$\tilde{z}(t) = \frac{dz}{dt} \quad \text{and} \quad u_\mu(0) \rightharpoonup z(0) \quad \text{weakly in } V \text{ as } \mu \to \infty.$$

Hereby, the initial condition u^0 is fulfilled since by assumption $u_\mu = u_{0\mu} \to u_0$. Hence $z(0) = y_0$.

Suppose now that $\varphi \in C^1([0, T])$ with $\varphi(T) = 0$. Now, put $\varphi_j = \varphi(t) w_j$ and multiply equation (11.65) by $\varphi(t)$ and make μ and j independent of each other by taking $m = \mu > j$. Hence,

$$\int_0^T \left[-(u'_\mu, \varphi'_j(t))_H + a(t; u_\mu(t), \varphi_j(t)) \right] dt = \int_0^T (f(t), \varphi_j(t))_H dt$$
$$+ (u_{1\mu}, \varphi_j(0))_H$$

and as $\mu \to \infty$ we have

$$\int_0^T \left[-(z'(t), \varphi'_j(t))_H + a(t; z(t), \varphi_j(t)) \right] dt = \int_0^T (f(t), \varphi_j(t))_H dt$$
$$+ (u_1, \varphi_j(0))_H. \qquad (11.76)$$

If $\varphi \in C_0^\infty([0, T])$, then

$$\frac{d^2}{dt^2}(z(t), w_j)_H \varphi(t) + a(t; z(t), w_j)\varphi(t) = (f(t), w_j)_H \varphi(t) \quad \forall j \in N,$$
$$(11.77)$$

where the differentiation is to be understood in the sense of distributions. Hence,

$$z''(t) + A(t)z(t) = f(t).$$

Using equation (11.76) and (11.77) we obtain

$$\varphi(0)(z'(0), w_j)_H = \varphi(0)(y_1, w_j)_H$$

for all w_j, that is, $z'(0) = u_1$. We conclude that $z(t)$ is a solution of the problem.

The proof of Theorem 11.8 is completed. ∎

Inequality (11.71) in the previous proof actually gives us a stronger regularity than stated in Theorem 11.8, as also mentioned in Remark 11.6. In Example 11.8 we derive from inequality (11.71) that

$$\|u_m(t)\|_V + \|u'_m(t)\|_H \leq K \tag{11.78}$$

holds for some constant K. This means that we have uniform bounds for $u_m(t) \in L^\infty(0,T;V)$ and $u'_m(t) \in L^\infty(0,T;H)$. Since L^1 is a separable Banach space, we conclude that

$$u_m(t) \rightharpoonup z(t) \quad \text{weakly* in } L^\infty(0,T;V),$$
$$u'_m(t) \rightharpoonup \tilde{z}(t) \quad \text{weakly* in } L^\infty(0,T;H).$$

From this we cannot conclude weak convergence as in the previous proof since L^1 is not a reflexive space. But weak* convergence in $L^\infty(0,T;V)$ implies weak convergence in $L^2(0,T;V)$ (since $L^\infty(]0,T[)$ is continuously imbedded in $L^2(]0,T[)$ and $L^2(]0,T[)$ is reflexive), so we derive from Example 11.8 that we can extract a weakly convergent subsequence in $L^2(0,T;V)$. Hence $u_m(0)$ converges weakly in V to $z(0)$. By assumption $u_m(0)$ converges strongly in V to u^0 and hence $z(0) = u^0$. In the same way as in the proof of Theorem 11.8 we see that the initial condition u^1 is fulfilled. We also see that $z(t)$ is the solution of the problem stated in Theorem 11.8, and that the differential equation is valued in the sense of V'-valued distributions. By this we extend the result of Theorem 11.8 to a continuous linear mapping of

$$L^2(0,T;H) \times V \times H \to L^\infty(0,T;V) \times L^\infty(0,T;H),$$

and we have thus justified Remark 11.6.

Example 11.8

Let $u \in C^1([0,T];V)$. We will, using the notation of Theorem 11.8, see that

$$\|u_m(t)\|_V + \|u'_m(t)\|_H \leq C \left[\|u_0\|_V + \|u_1\|_H + \int_0^T \|f(s)\|_H ds \right] = K,$$

where C and K are a constants.

Using the fact that V is continuously imbedded in H, we obtain directly from (11.71) that

$$\|u_m(t)\|_V^2 + \|u'(t)\|_H^2 \leq c_1 \left(\|u_{0m}\|_V^2 + \|u_{1m}\|_H^2\right)$$

$$+ c_1 \int_0^t \|u_m(s)\|_V + \|u'_m(s)\|_H^2 ds$$

$$+ c_1 \sup_{0 \leq t \leq T} \{\|u'_m(t)\|_H\} \int_0^t \|f(s)\|_H ds. \quad (11.79)$$

If we notice that

$$2 \sup_{0 \leq t \leq T} \{\|u'_m(t)\|_H\} \int_0^t \|f(s)\|_H ds \leq 2 \left(\int_0^t \|f(s)\|_H ds\right)^2$$

$$+ \frac{1}{2} \left(\sup_{0 \leq t \leq T} \{\|u'_m(t)\|_H\}\right)^2 \quad (11.80)$$

and define

$$N(t, u) = \sup_{0 \leq \tilde{t} \leq t} \left[\|u(\tilde{t})\|_V^2 + \|u'(\tilde{t})\|_H^2\right], \quad (11.81)$$

we see immediately that (11.79) can be written as

$$N(t, u) \leq C_1 N(0, u) + C_1 \int_0^t N(s, u) ds +$$

$$2 C_1 \left(\int_0^t \|f(s)\|_H ds\right)^2 \frac{1}{2} \left(\sup_{0 \leq t \leq T} \{\|u'_m(t)\|_H\}\right)^2$$

$$+ C_1 \frac{1}{2} \left(\sup_{0 \leq t \leq T} \{\|u'_m(t)\|_V\}\right)^2.$$

Hence,

$$N(t, u) \leq C_2 \left[N(0, u) + \int_0^t N(s, u) ds + \left(\int_0^t \|f(s)\|_H ds\right)^2\right]. \quad (11.82)$$

If we apply Gronwall's Lemma in (11.82) we obtain

$$N(t, u) \leq C_2 \left[N(0, u) + \left(\int_0^t \|f(s)\|_H ds\right)^2\right]. \quad (11.83)$$

If we first use $(a + b)^2 \leq 2(a^2 + b^2)$ for $a, b \geq 0$ in (11.83), and subsequently $(a^2 + b^2)^{\frac{1}{2}} \leq a + b$, we get the result, namely

$$\|u_m(t)\|_V + \|u'_m(t)\|_H \leq C \left[\|u_0\|_V + \|u_1\|_H + \int_0^T \|f(s)\|_H ds\right] = K,$$

where C is a constant.   □

As mentioned in Remark 11.6, we can extend the result of Theorem 11.8 as in Theorem 11.9 below.

THEOREM 11.9

Subject to the assumptions in Section 11.7, the problem

$$\begin{cases} u'' + Au = f \text{ in } \Omega \times]0, T[, \\ u(x,0) = u^0, \ u'(x,0) = u^1 \text{ in } \Omega, \end{cases} \tag{11.84}$$

where

$$\begin{cases} F \in L^2(0,T;H), \\ u^0 \in V, \ u^1 \in H \end{cases} \tag{11.85}$$

has a unique solution u. The map

$$\{F, u_0, u_1\} \to \left\{ u, \frac{\partial u}{\partial t} \right\} \tag{11.86}$$

is continuous and linear from

$$L^2(0,T;H) \times V \times H \to C(0,T;V) \times C(0,T;H). \tag{11.87}$$

Before we prove Theorem 11.9 we need some preliminary lemmas.

In the following we will use weakly continuous functions which are defined in Definition 11.15 below.

DEFINITION 11.15   [Weak Continuity] *Let Y be a Banach space. By $C_w(0,T;Y)$ we denote the space of functions $f \in L^\infty(0,T;Y)$ which are weakly continuous, that is, $\langle f(t), y' \rangle$ is continuous on $[0,T]$, $\forall y' \in Y'$.*

LEMMA 11.10

Let X and Y be Banach spaces such that $X \subset Y$ with continuous injection. Moreover, let X be a reflexive space. Then

$$L^\infty(0,T;X) \cap C_w(0,T;Y) = C_w(0,T;X).$$

PROOF   We show that

$$\|f(t)\|_X \le C, \tag{11.88}$$

where C is constant. We can assume without loss, generally, that $f \in L^\infty(R; X)$.

Let $\varrho_\varepsilon \in C_0^\infty(R)$ be a mollifier with $\int_R \varrho_\varepsilon dt = 1$, which vanishes for $|x| \geq \varepsilon$. Then, since $f \in L^\infty(R; X)$, we have that

$$\|f * \varrho_\varepsilon\|_X \leq M$$

where M is a constant. Hence for a subsequence (still denoted by ε), we have

$$f * \varrho_\varepsilon \rightharpoonup \tilde{f}(t) \quad \text{weakly in } X.$$

By assumption, $\langle f(t), y' \rangle$ is continuous in t for every $y' \in Y'$. Hence, by using the properties of the ϱ_ε we have

$$\langle f(t) * \varrho_\varepsilon - f(t), y' \rangle = \varrho_\varepsilon * \langle f(t), y' \rangle - \langle f(t), y' \rangle \to 0 \quad \text{as } \varepsilon \to 0.$$

From this we have

$$f * \varrho_\varepsilon \rightharpoonup f(t) \quad \text{weakly in } Y.$$

Hence $\tilde{f}(t) = f(t)$, and since $\tilde{f}(t)$ bounded in X, we obtain (11.88).

We now show that $f \in C_w(0, T; X)$. Let $s \to t$. Since $f(s)$ bounded in X, we can extract a subsequence (still denoted $f(s)$) such that $f(s) \rightharpoonup \chi$ weakly in X as $s \to t$. Now since $X \subset Y$ and $f(s) \rightharpoonup f(t)$, we conclude that $\chi = f(t)$. ∎

LEMMA 11.11
Let u be the solution from Theorem 11.8. Then

$$u \in C_w(0, T; V) \quad \text{and} \quad u' \in C_w(0, T; H) \tag{11.89}$$

(after a possible modification on a set with measure zero).

PROOF   From Remark 11.6 we have that $u \in L^\infty(0, T; V)$ and $u' \in L^\infty(0, T; H)$. Then $u \in C([0, T]; H)$. Hence $u \in C_w(0, T; V)$ by Lemma 11.10 with $X = V$ and $Y = H$.

Since $u'' = f - Au \in L^2(0, T; V')$ hence $u' \in C([0, T]; V')$. Then $u' \in C_w(0, T; H)$ by another application of Lemma 11.10. ∎

LEMMA 11.12
Let u be the solution from Theorem 11.8. Then for all t

$$a(t; u(t), u(t)) + \|u'(t)\|_H^2 = a(0; u_0, u_0) + \|u'(0)\|_H^2$$
$$+ \int_0^t a'(s; u, u) ds$$
$$+ 2 \int_0^t (f, u')_H ds. \tag{11.90}$$

PROOF   See Lions and Magenes [19], Vol. I, p. 276.   ∎

PROOF OF THEOREM 11.9
From Lemma 11.11 we have

$$u \in C_w(0, T; V), \quad \text{and } u' \in C_w(0, T; H).$$

If we define $\varphi(t)$ by

$$\varphi(t) = a(t; u(t), u(t)) + \|u'(t)\|_H^2,$$

we get from Lemma 11.12 that $\varphi(t)$ is continuous on $[0, T]$. Moreover, we know that $u(t) \in C([0, T]; H)$.
Now let $s \to t$ and define ξ_n by

$$\xi_n = \|u'(s) - u'(t)\|_H^2 + a(s; u(s) - u(t), u(s) - u(t)).$$

We have

$$\xi_n = \varphi(s) + \varphi(t) + 2\left[a(s; u(t), u(t)) - a(t; u(t), u(t))\right]$$
$$-2(u'(s), u(t))_H - a(t, u(s), u(s)) - 2\left[a(s; u(s), u(t)) - a(t; u(s), u(t))\right].$$

By exploiting the weak continuity of $u(t)$ and $u'(t)$ and the continuity of $\varphi(t)$, we have first

$$|a(s; u(t), u(t)) - a(t; u(t), u(t))| < c_1|t - s|,$$
$$|a(s; u(t), u(s)) - a(t; u(t), u(s))| < c_2|t - s|,$$

and secondly

$$\xi_n \to 2\varphi(t) - 2\left(\|u'(t)\|_H^2 + a(t; u(s), u(t))\right) = 0 \text{ as } \varepsilon \to 0.$$

Since

$$\xi_n \geq \|u'(s) - u'(t)\|_H^2 + \alpha\|u(s) - u(t)\|_V^2,$$

the proof is completed.

11.8.2   An Abstract Regularity Theorem

In this section we present a regularity theorem that is in principle built upon Theorem 11.8 in combination with the compatibility conditions on the initial data. The idea of the proof is to repeatedly write up the problem of Theorem 11.8 by using substitution, and the compatibility conditions as initial data.

THEOREM 11.13 (An Abstract Regularity Theorem)

We consider the problem

$$u''(t) + A(t)u(t) = f(t) \tag{11.91}$$

with the initial conditions

$$u(0) = u^0 \ , \ u'(0) = u^1. \tag{11.92}$$

In addition to the assumptions in Section 11.7, assume that

$$f \in H^{k-1}(0,T;H), \ k \geq 1,$$

and that we have the compatibility conditions

$$\frac{d^j u(0)}{dt^j} \in V, \quad j = 0, ..., k-1, \quad \frac{d^k u(0)}{dt^k} \in H.$$

Then the solution u of Problem (11.91)-(11.92) satisfies

$$u \in H^{k-1}(0,T;V) \ , \ \frac{d^k u(t)}{dt^k} \in L^2(0,T;H) \ , \ \frac{d^{k+1} u(t)}{dt^{k+1}} \in L^2(0,T;V').$$

REMARK 11.7   The compatibility conditions can also be written as

$$\frac{\partial^{2n-1} u(0)}{\partial t^{2n-1}} = f^{(2n-3)}(0) - A f^{(2n-5)}(0) + \cdots + (-1)^{n-2} A^{n-2} f'(0)$$
$$+ (-1)^{n-1} A^{n-1} u_1,$$

$$\frac{\partial^{2n} u(0)}{\partial t^{2n}} = f^{(2n-2)}(0) - A f^{(2n-4)}(0) + \cdots + (-1)^{n-1} A^{n-1} f'(0)$$
$$+ (-1)^n A^n u_0.$$

∎

PROOF   See Wloka [31] p. 443.   ∎

11.9   Transposition

The method of transposition is frequently used because it provides a useful tool for finding unique weak solutions. The basic idea behind the method of transposition is to exploit a well known regularity theorem to obtain a regularity result of another problem. The procedure, loosely speaking, corresponds to finding the adjoint operator. Therefore, it seems reasonable that the solvability of the problem we wish to solve somehow dictates which well known regularity theorems can be used in the transposition procedure. With this in mind, we will illustrate the underlying idea of transposition by solving problem (11.93) stated below by transposition.

$$\begin{cases} Au + u'' = f \in L^2(0,T;H), \\ u(0) = u^0 \quad, \; u'(0) = u^1, \end{cases} \tag{11.93}$$

where $u^0 \in H$ and $u^1 \in V'$. The first step is the choice of a regularity theorem and our choice here is to consider the following system:

$$\begin{cases} Av + v'' = \varphi \in L^2(0,T;H), \\ v(T) = v'(T) = 0. \end{cases} \tag{11.94}$$

According to Theorem 11.9 we have

$$v \in C^0([0,T];V) \text{ and } v' \in C^0([0,T];H). \tag{11.95}$$

By considering the space of solutions of the problem (11.94) as f runs through $L^2(0,T;H)$, we obtain an isomorphism, as we shall see. For this purpose we introduce the definition below.

DEFINITION 11.16   *Let v be a solution of the problem (11.94) and let $\varphi \in L^2(0,T;H)$. Then we define the space X as*

$$X = \left\{ \text{ The space of solutions } v \text{ as } \varphi \text{ runs through } L^2(0,T;H) \right\}.$$

Providing X with the topology carried over by the mapping $\varphi \to v$, we are ensured that $A + \frac{d^2}{dt^2}$ is an isomorphism of X onto $L^2(0,T;H)$. From this conclusion we deduce the following theorem.

THEOREM 11.14
Assume that the hypothesis of Section 11.9 is fulfilled. Let L be a continuous antilinear form on U. Then there exists a unique $u \in L^2(0,T;H)$ such that

$$(u, Av + v'')_{L^2(0,T;H)} = (Lu, v)_{X',X} \quad \forall v \in X. \tag{11.96}$$

The last step is to choose L such that u satisfies the system (11.93). We take

$$(Lu, v)_{U',U} = \int_{]0,T[} (f,v)dt + (u_1, v(0)) - (u_0, v'(0)),$$

giving suitable meaning to the various scalar products occuring in this formula. Then by an integration by parts, it is seen that u formally satisfies

$$\begin{cases} Au + u'' = f \in L^2(0,T;H), \\ u(0) = u^0, \ u'(0) = u^1, \end{cases} \tag{11.97}$$

where $u^0 \in H$ and $u^1 \in V'$. We have thus obtained the result.

11.9.1 Transposition of Nonhomogeneous Boundary Value Problems

In the section to come we will present a number of regularity theorems of nonhomogeneous boundary value problems of the type stated in (11.98). Some of the proofs of these theorems are built upon transposition, and we will therefore give a description of the method used on the system below.

$$\begin{cases} Au + u'' = f & \text{in } \Omega \times]0, T[, \\ B_j u = g_j & \text{on } \Gamma \times]0, T[, \ 0 \leq j \leq m - 1, \\ u(0) = u^0, \ u'(0) = u^1. \end{cases} \tag{11.98}$$

The procedure in solving problem (11.98) resembles the one mentioned in the previous section. Hence, the first step is the choice of a system we can utilize to solve the problem (11.98). Our choice is the system below.

$$\begin{cases} Av + v'' = \psi & \text{in } \Omega \times]0, T[, \\ C_j v = 0 & \text{on } \Gamma \times]0, T[, \ 0 \leq j \leq m - 1, \\ v(T) = v'(T) = 0. \end{cases} \tag{11.99}$$

Here again, a suitable regularity theorem for the system (11.99) will be used. By suitable, we mean that if we change the right-hand side ψ in problem (11.99), we may have to change $G = \{f, \{g_j\}, u_0, u_1\}$ in problem (11.98), and vice versa. Following the idea of the previous section, the reason for this is that we are searching for a unique solution u of

$$(u, Av + v'') = \int_{]0,T[} (f,v)dt + \sum_{j=0}^{m-1} (g_j, T_j v) \tag{11.100}$$

$$+ (u_1, v(0)) - (u_0, v')$$
$$+ (u(T), v'(T)) - (u'(T), v(T))$$

and have to give suitable meaning to the various scalar products in this formula. Hereafter, if we consider Green's formula, we see that the signs in (11.100) are chosen such that u formally satisfies Problem (11.98).

In order to keep the presentation simple, we present the result of transposition without a fixed regularity theorem of Problem (11.99). Therefore, let $\psi \in U$, say, and consider all the solutions of Problem (11.99) as ψ traverses U. Thereby, we get a similar isomorphism as in the previous section, and we can present the following definition.

DEFINITION 11.17   A weak solution
Let $\psi \in U$ and define the space X as:

$X = \{$ the space of solutions as ψ runs through the space $U\}$.

We now define the weak solution of the system (11.98) to be the unique solution u which satisfies the following equation (11.101).

$$(u, Av + v'') = \int_{]0,T[} (f, v)dt + \sum_{j=0}^{m-1} (g_j, T_j v) \qquad (11.101)$$
$$+(u_1, v(0)) - (u_0, v'(0))$$
$$+(u(T), v'(T)) - (u'(T), v(T)), \quad \forall v \in X,$$

where $v(0)$ and $v'(0)$ in (11.101) are the initial data for the system (11.99).

REMARK 11.8   In this particular case, the solution v of a specific regularity theorem has to coincide with $G = \{f, \{g_j\}, u_0, u_1\}$ according to the scalar products on the right-hand side of (11.101).   ∎

REMARK 11.9   In all our cases, the assumptions in Section 11.7 will ensure the existence of a unique solution.   ∎

For a comprehensive study of transposition we refer to Lions and Magenes Vol. I, Vol. II [19].

Chapter 12

Regularity of Hyperbolic Mixed Problems

The aim of this chapter is to investigate the regularity question of hyperbolic mixed problems in the Dirichlet case in order to be able to apply the *Hilbert Uniqueness Method* introduced in the next chapter. But since this chapter outlines the variational methods for the study of hyperbolic mixed problems, it can be read independently.

Let $A(x, t)$ denote a symmetric linear elliptic operator of order $2m$. The problem to be studied is

$$\begin{cases} \Phi'' + A(x,t)\Phi = f \text{ in } \Omega \times]0, T[, \\ \Phi(x, 0) = \Phi^0, \ \Phi'(x, 0) = \Phi^1 \text{ in } \Omega, \\ B_j \Phi = g_j, \ 0 \le j \le m - 1, \quad \text{on } \Gamma \times]0, T[. \end{cases} \qquad (12.1)$$

Under various assumptions on the regularity of F, g_j, Φ^0, and Φ^1, we seek optimal regularity results of the solution Φ and the traces thereof. These solvability results for hyperbolic mixed problems are highly applicable in the theory of *boundary control of systems of evolution type*, as will become evident in the ensuing chapter.

The regularity theorems can be divided into two categories, the "classical" and the "newer" regularity results, respectively. The classical regularity results can be found in Lions and Magenes [19], and the newer regularity results are based on results obtained in the early eighties and can be found in a collected form in Lasiecka, Lions, and Triggiani [16].

We first state all the results and comment on them briefly in Sections 12.1 and 12.1.3. The proofs will subsequently follow. The inclusion of the proofs of *all* the results presented is beyond the scope of this book. In the classical case we have already proved, in Section 11.8, Chapter 11, a result which plays a vital role in obtaining the newer regularity results. The proofs of all the newer results are presented in Section 12.2.

12.1 Solvability Results

We present in this section a selected sample of solvability results for non-homogeneous boundary value problems.

12.1.1 Classical Solvability Results

We begin by presenting some classical solvability theorems. The proofs of these results can be found in Lions and Magenes [19] Vol. I, Chapter 3 and Vol. II, Chapters 4 and 5.

Interior Regularity

Let A be an elliptic, symmetric operator of order $2m$ with sufficiently smooth coefficients as introduced in Section 11.1, Chapter 11. Let B_j denote the boundary operators as introduced in Section 11.3, Chapter 11. In Theorem 12.1 below, the boundary conditions are contained in the space V.

THEOREM 12.1
Subject to the assumptions in Section 11.7, the problem

$$\begin{cases} u'' + A(x,t)u = f \text{ in } \Omega \times]0, T[, \\ u(x,0) = u^0, \ u'(x,0) = u^1 \text{ in } \Omega, \end{cases} \tag{12.2}$$

where

$$\begin{cases} f \in L^1(0,T;H), \\ u^0 \in V, \ u^1 \in H, \end{cases} \tag{12.3}$$

has a unique solution u. The map

$$\{f, u_0, u_1\} \rightarrow \left\{ u, \frac{\partial u}{\partial t} \right\} \tag{12.4}$$

is continuous and linear from

$$L^1(0,T;H) \times V \times H \rightarrow C(0,T;V) \times C(0,T;H). \tag{12.5}$$

PROOF
We have proven this theorem in Section 11.8, Chapter 11. ∎

We now state a theorem that gives us an interior regularity result when the boundary conditions are homogeneous (i.e., $B_j u = g_j = 0$).

THEOREM 12.2
Consider the problem

$$\begin{cases} u'' + A(x,t)u = f & \text{in } \Omega \times]0, T[, \\ B_j u = 0, \ \ 0 \leq j \leq m-1, & \text{on } \Gamma \times]0, T[, \\ u(x,0) = u^0, \ \ u'(x,0) = u^1 & \text{on } \Omega. \end{cases} \tag{12.6}$$

Assume that

$$\begin{cases} f \in L^2(\Omega \times]0, T[), \\ f' \in L^2(\Omega \times]0, T[) \end{cases} \tag{12.7}$$

and

$$\begin{cases} u^0 \in H^{2m}(\Omega), \\ u^1 \in H^m(\Omega). \end{cases} \tag{12.8}$$

Then the solution of (12.6), which exists and is unique, satisfies

$$\begin{cases} u \in L^2(0, T; H^{2m}(\Omega)), \\ u'' \in L^2(\Omega \times]0, T[). \end{cases} \tag{12.9}$$

PROOF   See Lions and Magenes [19], Vol. II, pp. 96-99.   ∎

Reduction to the Homogeneous Case

Theorem 12.2 gives a result when the boundary conditions are homogeneous. We are, however, mainly interested in the case $g_j \neq 0$. In order to solve a nonhomogeneous problem, we can reduce it to a homogeneous one.
Consider the following problem.

$$\begin{cases} u'' + A(x,t)u = f, \\ B_{ju} = g_j, \ \ 0 \leq j \leq m-1 & \text{on } \Gamma \times]0, T[, \\ u(x,0) = u^0, \ \ u'(x,0) = u^1 & \text{on } \Omega. \end{cases} \tag{12.10}$$

Now, assume that there exists a w solving the following system:

$$w(x,0) = u^0, \ \ w'(x,0) = u^1.$$

REMARK 12.1   Notice that w is unique.   ∎

If we set

$$\theta = u - w \qquad (12.11)$$

we obtain the system:

$$\begin{cases} \theta'' + A(x,t)\theta = f - \Psi, \\ B_j\theta = 0, \ \ 0 \leq j \leq m-1 \ \ \text{on } \Gamma \times]0, T[, \\ \theta(x,0) = \theta'(x,0) = 0. \end{cases} \qquad (12.12)$$

If we are able to make the reduction above, we can apply Theorem 12.2 to problem (12.12) and hereafter use (12.11). Notice that the method of reduction also applies for the backward problem.

In order to obtain regularity results, we are obliged to characterize Ψ in a "suitable" way. The so-called compatibility conditions play an important role here. Roughly speaking, the compatibility conditions express that the boundary data and the initial data have to coincide at the initial stage. The compatibility conditions state that not all the initial conditions and boundary data can be chosen independently of each other if we wish to have consistency in the problem. We refer the reader to Lions and Magenes [19], Vol. II, Chapters 4 and 5, for a more comprehensive presentation.

We now state a theorem that holds in the nonhomogeneous case in which we also demand that some compatibility conditions are fulfilled.

THEOREM 12.3
Consider the problem

$$\begin{cases} u'' + A(x,t)u = f \ \ \text{in } \Omega \times]0, T[, \\ \frac{\partial^j u}{\partial \nu^j} = g_j, \ \ 0 \leq j \leq m-1 \ \ \text{on } \Gamma \times]0, T[, \\ u(x,0) = u^0, \ \ u'(x,0) = u^1 \ \text{in } \Omega. \end{cases} \qquad (12.13)$$

Let

$$\begin{cases} f \in L^2(\Omega \times]0, T[), \\ g_j \in H^{2m-j-\frac{1}{2},(2m-j-\frac{1}{2})/m}(\Gamma \times]0, T[), \ \ 0 \leq j \leq m-1, \\ u^0 \in H^{\frac{3m}{2}}(\Omega), \ \ u^1 \in H^{\frac{m}{2}}(\Omega) \end{cases} \qquad (12.14)$$

with the following compatibility conditions:
there exists a $w \in H^{2m,2}(\Omega \times]0, T[)$ *satisfying*

$$\frac{\partial^j w}{\partial \nu^j} = g_j, \quad 0 \leq j \leq m - 1, \quad w(x, 0) = u^0, \quad w'(x, 0) = u^1. \qquad (12.15)$$

Then Problem (12.13) - (12.15) has a unique solution such that

$$u \in H^{m,1}(\Omega \times]0, T[). \qquad (12.16)$$

(Recall that $H^{r,s}(\Omega \times]0, T[) = H^0(0, T; H^r(\Omega)) \cap H^s(0, T; H^0(\Omega)).)$

PROOF
See Lions and Magenes [19], Vol. II, p. 105.
∎

Regularity of the Normal Derivative
The regularity of the normal derivative plays a vital role in the theory of boundary controllability that will be presented in the next chapter. The following classical result gives us a result for the regularity of the normal derivative given a certain interior regularity.

THEOREM 12.4
For $u \in H^{r,s}(\Omega \times]0, T[)$ *with* $r > 0, s \geq 0$ *and* $j \in N \cup \{0\}$, *we have*

$$\frac{\partial u}{\partial \nu} \in H^{\mu_j, \eta_j}(\Gamma \times]0, T[) \quad \text{if } j < r - \frac{1}{2}, \qquad (12.17)$$

where

$$\frac{\mu_j}{r} = \frac{\eta_j}{s} = \frac{r - j - \frac{1}{2}}{r} \quad (\eta_j = 0 \text{ if } s = 0) \qquad (12.18)$$

and $u \to \frac{\partial^j u}{\partial \nu^j}$ *are continuous linear mappings of*

$$H^{r,s}(\Omega \times]0, T[) \to H^{\mu_j, \eta_j}(\Gamma \times]0, T[).$$

PROOF   See Lions and Magenes [19], Vol. II, p. 9.   ∎

Example 12.1
Assume that a function Φ has the following interior regularity

$$\Phi \in H^{2,2}(\Omega \times]0, T[). \qquad (12.19)$$

We can use Theorem 12.4 above to obtain a regularity result for the normal derivative. Using the notation of the theorem we have

$$r = s = 2.$$

In order to get the regularity of the normal derivative, we wish to investigate the case $j = 1$. In this case we get (cf. equation (12.18))

$$\mu_j = \nu_j = \frac{1}{2}. \tag{12.20}$$

According to the classical results for the normal derivative, we have therefore the following regularity of the normal derivative

$$\frac{\partial \Phi}{\partial \nu} \in H^{\frac{1}{2}, \frac{1}{2}}(\Gamma \times]0, T[). \tag{12.21}$$

We shall see (cf. Theorem 1 below) that this result can actually be improved.   ∏

12.1.2   Newer Solvability Theorems

We now state some of the more recent solvability theorems. The proofs of all the theorems stated here can be found in Section 12.2. We first consider the canonical situation where the system's evolution is governed by the classical wave equation. The problem we wish to solve is

$$\begin{cases} \Phi'' - \Delta \Phi = F \text{ in } \Omega \times]0, T[, \\ \Phi(x,0) = \Phi^0, \ \Phi'(x,0) = \Phi^1 \text{ in } \Omega, \\ \Phi = g \text{ on } \Gamma \times]0, T[. \end{cases} \tag{12.22}$$

Theorem 1 *Consider the problem (12.22). The hypotheses are:*

$$\begin{cases} F \in L^1(0, T; L^2(\Omega)), \\ \Phi^0 \in H^1(\Omega), \ \Phi^1 \in L^2(\Omega), \\ g \in H^1(\Gamma \times]0, T[) \end{cases} \tag{12.23}$$

with the compatibility conditions

$$g|_{t=0} = \Phi^0|_\Gamma. \tag{12.24}$$

Then the unique solution Φ of (12.22) satisfies

$$\begin{cases} \Phi \in C([0, T]; H^1(\Omega)), \\ \Phi' \in C([0, T]; L^2(\Omega)) \end{cases} \tag{12.25}$$

and

$$\frac{\partial \Phi}{\partial \nu} \in L^2(\Gamma \times]0, T[). \tag{12.26}$$

REMARK 12.2   In the case of Theorem 1, the mapping

$$\{F, \Phi^0, \Phi^1, g\} \rightarrow \left\{\Phi, \Phi', \frac{\partial \Phi}{\partial \nu}\right\} \tag{12.27}$$

from a subspace of

$$L^1(0, T, L^2(\Omega)) \times H^1(\Omega) \times L^2(\Omega) \times H^1(\Gamma \times]0, T[) \text{ into}$$
$$C([0, T]; H^1(\Omega)) \times C([0, T]; L^2(\Omega)) \times L^2(\Gamma \times]0, T[)$$

is continuous.   ∎

A similar remark applies to all the following results.

Theorem 2 *Consider the problem (12.22). The hypotheses are:*

$$\begin{cases} F \in L^1(0, T; H^1(\Omega)), \\ F' \in L^1(0, T; L^2(\Omega)), \\ \Phi^0 \in H^2(\Omega), \ \Phi^1 \in H^1(\Omega), \\ g \in H^2(\Gamma \times]0, T[) \end{cases} \tag{12.28}$$

with the compatibility conditions

$$g|_{t=0} = \Phi^0|_\Gamma, \ g'|_{t=0} = \Phi^1|_\Gamma. \tag{12.29}$$

Then the unique solution Φ *of (12.22) satisfies*

$$\begin{cases} \Phi \in C([0, T]; H^2(\Omega)), \\ \Phi' \in C([0, T]; H^1(\Omega)), \\ \Phi'' \in C([0, T]; L^2(\Omega)) \end{cases} \tag{12.30}$$

and

$$\frac{\partial \Phi}{\partial \nu} \in H^1(\Gamma \times]0, T[). \tag{12.31}$$

Weak Solution
If we take our data $\{F, g, \Phi^0, \Phi^1\}$ weaker than in the previous section, we will not expect the same degree of regularity of the solution. In order to obtain this weaker solution, transposition (cf. Section 11.9) will be used.

Theorem 3 *Consider the problem (12.22). The hypotheses are:*

$$\begin{cases} F \in L^1(0, T; H^{-1}(\Omega)), \\ \Phi^0 \in L^2(\Omega), \ \Phi^1 \in H^{-1}(\Omega), \\ g \in L^2(\Gamma \times]0, T[) \end{cases} \tag{12.32}$$

without any compatibility conditions this time. Then the unique solution Φ
of (12.22) satisfies

$$\begin{cases} \Phi \in C([0,T]; L^2(\Omega)), \\ \Phi' \in C([0,T]; H^{-1}(\Omega)) \end{cases} \tag{12.33}$$

and

$$\frac{\partial \Phi}{\partial \nu} \in H^{-1}(\Gamma \times]0, T[). \tag{12.34}$$

REMARK 12.3   Theorems 1, 2, and 3 all remain true if we consider
the following system

$$\begin{cases} \Phi'' + A(x,t)\Phi = F \text{ in } \Omega \times]0, T[, \\ \Phi(x,0) = \Phi^0, \Phi'(x,0) = \Phi^1 \text{ in } \Omega, \\ \Phi = g \text{ on } \Gamma \times]0, T[, \end{cases} \tag{12.35}$$

where $A(x,t)$ is a second order symmetric elliptic operator and

$$A = -\frac{\partial}{\partial x_i}\left(a_{ij}(x,t)\frac{\partial}{\partial x_j}\right),$$

$$a_{ij}(x,t), a'_{ij}(x,t) \text{ and } \frac{\partial a_{ij}}{\partial x_k} \in L^\infty(\Omega \times]0, T[).$$

∎

12.1.3   A Discussion of the Regularity Results

Some remarks on the regularity results stated in the previous section are
now in order. Let us consider a system described by the classical wave
equation,

$$\begin{cases} \Phi'' - \Delta\Phi = F \text{ in } \Omega \times]0, T[, \\ \Phi(x,0) = \Phi^0, \quad \Phi'(x,0) = \Phi^1 \text{ in } \Omega, \\ \Phi = g \text{ on } \Gamma \times]0, T[. \end{cases} \tag{12.36}$$

In Theorem 1 the following interior regularity of the solution to the problem
(12.36) under appropriate assumptions on the initial and boundary data is
obtained:

$$\Phi \in C([0,T]; H^1(\Omega)) \cap C^1([0,T]; L^2(\Omega)). \tag{12.37}$$

Using this interior regularity, we have achieved the following regularity for
the normal derivative:

$$\frac{\partial \Phi}{\partial \nu} \in L^2(\Gamma \times]0, T[). \tag{12.38}$$

This boundary regularity does *not* follow from the interior regularity, but is an independent regularity result.

In the proof of the regularity of the normal derivative we use the multiplier $h \cdot \nabla \Phi$, where h is a smooth vector field and $h \cdot \nu = 1$ on Γ. After various integration by parts we obtain

$$\left\| \frac{\partial \Phi}{\partial \nu} \right\|_{L^2(\Gamma \times]0, T[)} \quad \text{in terms of } \{\Phi^0, \Phi^1, g, F\}.$$

As we will see, this technique also proves to be useful in the exact controllability problems with Dirichlet boundary control on a part of the boundary $\Gamma_0 \subset \Gamma$ that we present later.

In the proofs to come, the full strength of the assumption $g \in H^k(\Gamma \times]0, T[)$ in Theorems 1, 2, and 3 is used only to obtain the regularity of the normal derivative. Let us elaborate a little bit on this by considering the case of Theorem 1. Since $g \in H^1(\Gamma \times]0, T[)$, it satisfies for an even stronger reason (cf. The Continuous Derivatives Theorem)

$$g \in C([0, T]; [H^1(\Gamma), H^0(\Gamma)]_{\frac{1}{2}}) = C([0, T]; H^{\frac{1}{2}}(\Gamma)). \tag{12.39}$$

This is exactly the condition that is sufficient to obtain the interior regularity of the solution in the proof of Theorem 1. (See also Remark 12.5, Section 12.2.)

Similarly, in the case of Theorem 2 we have $g \in H^2(\Gamma \times]0, T[)$. Once again, according to The Continuous Derivatives Theorem, we have

$$\begin{cases} g \in C([0, T]; [H^2(\Gamma), H^0(\Gamma)]_{\frac{1}{4}}) = C([0, T]; H^{\frac{3}{2}}(\Gamma)), \\ g' \in C([0, T]; [H^2(\Gamma), H^0(\Gamma)]_{\frac{3}{4}}) = C([0, T]; H^{\frac{1}{2}}(\Gamma)). \end{cases}$$

The above conditions are sufficient to obtain the interior regularity.

In the theory of exact controllability in the next chapter, the most interesting problems are those problems that allow us to consider Dirichlet boundary inputs that are in $L^2(\Gamma \times]0, T[)$. A relevant question that can therefore be posed is whether there has been any improvement in the interior regularity of the solution in the newer regularity theorems. We can only use the classical regularity result for Dirichlet boundary data in $L^2(\Gamma \times]0, T[)$ if the initial data are assumed to be zero. Even then, the solutions that we obtain lie in a somewhat "unreasonable" space. A comparison with the regularity achieved in Theorem 3 shows that this theorem, which allows us to use quite "weak" initial data is a clear improvement on the classical result.

Moreover, the interior regularity results for Φ and Φ' obtained in Theorem 3 are optimal in the sense that they coincide with the results available in the one-dimensional case (i.e., $n = 1$) where we can write an explicit formula for the solution Φ. This explicit solution in the one-dimensional case can be found in Lasiecka and Triggiani [17], Example 5.1, p. 52.

12.2   Proofs of the Newer Regularity Theorems

In this section we prove the newer regularity theorem stated in Section 12.1. The proofs are based on the proofs that can be found in Lasiecka, Lions, and Triggiani [16]. We rely upon multiplier techniques and energy estimates in our proofs. Recall that the problem we consider is

$$\begin{cases} \Phi'' - \Delta\Phi = F \text{ in } \Omega\times]0,T[, \\ \Phi(x,0) = \Phi^0, \ \Phi'(x,0) = \Phi^1 \text{ in } \Omega, \\ \Phi = g \text{ on } \Gamma\times]0,T[. \end{cases} \qquad (12.40)$$

First, we state and prove some auxiliary lemmas which will serve as requisites in the proofs of the theorems in section 12.1.2.

REMARK 12.4   In all the proofs, the continuity condition $C([0,T];X)$ will be replaced by $L^\infty(0,T;X)$. ■

LEMMA 12.5
Let ψ be given in $L^1(0,T;L^2(\Omega))$ and let φ be the solution of

$$\begin{cases} \varphi'' - \Delta\varphi = \psi \text{ in } \Omega\times]0,T[, \\ \varphi(x,0) = \varphi'(x,0) = 0 \text{ in } \Omega, \\ \varphi = 0 \text{ on } \Gamma\times]0,T[. \end{cases} \qquad (12.41)$$

Then we have

$$\varphi \in L^\infty(0,T,H^1(\Omega)), \quad \varphi' \in L^\infty(0,T,L^2(\Omega)) \qquad (12.42)$$

and the mapping $\psi \in L^1(0,T;L^2(\Omega)) \to \varphi$ is continuous.
We also have

$$\frac{\partial\varphi}{\partial\nu} \in L^2(\Gamma\times]0,T[) \qquad (12.43)$$

with the mapping $\psi \to \frac{\partial\varphi}{\partial\nu}$ being continuous from $L^1(0,T;L^2(\Omega)) \to L^2(\Gamma\times]0,T[$

PROOF   STEP 1
(12.42) follows directly from Theorem 11.8 (cf. Section 11.8).
STEP 2
In order to show (12.43) it is sufficient to prove that

$$\left\| \sqrt{T-t}\frac{\partial\varphi}{\partial\nu} \right\|_{L^2(\Gamma\times]0,T[)} \leq \|\psi\|_{L^1(0,T;L^2(\Omega))}$$

when ψ is smooth since T does not play a particular role. Hereafter, we extend by continuity.

Let $\nu = \{\nu_1, ..., \nu_n\}$ be the unit outward normal to Γ. Since Γ is of class C^2, we can always pick an extension $h_k \in C^1(\overline{\Omega})$, $h_k = \nu_k$ on Γ, of ν and define $h = \{h_1, \ldots, h_k\}$. Now if β is arbitrary in $H^1(\Omega)$, then $div(\beta h) = h \cdot \nabla\beta + \beta div(h)$. From this and the fact that $h \cdot \nu = 1$ on Γ, the divergence theorem gives

$$\int_\Omega h \cdot \nabla\beta dx = \int_\Gamma \beta d\Gamma - \int_\Omega \beta div(h)dx. \tag{12.44}$$

(12.44) will frequently be used in the forthcoming because $(T - t)h \cdot \nabla\varphi$ will be chosen as a multiplier below.
From the interior regularity (12.42) we know that

$$\psi \in L^1(0, T; L^2(\Omega)) \to \varphi, \quad \nabla\varphi, \quad \varphi' \in L^\infty(0, T; L^2(\Omega)) \tag{12.45}$$

is a continuous linear mapping. For simplicity, we shall write $O(\|\psi\|^2)$ for any expression bounded from above by $c\|\psi\|^2_{L^1(0,T;L^2(\Omega))}$ where c is a suitable constant.

STEP 3

If we multiply (12.41) by $(T - t)h \cdot \nabla\varphi$ we have

$$\int_{\Omega\times]0,T[} \varphi''(T - t)(h \cdot \nabla \varphi)dxdt - \int_{\Omega\times]0,T[} (T - t)\Delta\varphi(h \cdot \nabla\varphi)dxdt$$

$$= \int_{\Omega\times]0,T[} (T - t)\psi(h \cdot \nabla\varphi)dxdt. \tag{12.46}$$

Hence, for the first term on the L.H.S. we obtain by an integration by parts in t that

$$\int_{\Omega\times]0,T[} \varphi''(T - t)(h \cdot \nabla\varphi)dxdt = -T \int_\Omega \varphi'(0)(h \cdot \nabla\varphi^0)dx$$

$$+ \int_{\Omega\times]0,T[} \varphi'(h \cdot \nabla\varphi)dxdt$$

$$- \int_{\Omega\times]0,T[} \frac{T - t}{2}h \cdot \nabla(\varphi')^2 dxdt$$

$$\tag{12.47}$$

which equals

$$-T\int_\Omega \varphi^1(h\cdot\nabla\varphi^0)dx + \int_{\Omega\times]0,T[}\varphi'(h\cdot\nabla\varphi)dxdt$$

$$-\int_{\Gamma\times]0,T[}\frac{T-t}{2}(\varphi')^2 d\Gamma dt$$

$$+\int_{\Omega\times]0,T[}\frac{T-t}{2}(\varphi')^2 div\,(h)\,dxdt \qquad (12.48)$$

$$= O(\|\psi\|^2),$$

where we have used (12.45) and the fact that the first and the third integral above are zero because of the conditions in (12.41). Now since $h\cdot\nabla\varphi = \nu\cdot\nabla\varphi = \frac{\partial\varphi}{\partial\nu}$ on $\Gamma\times]0,T[$, we get for the second term, by Green's formula on the L.H.S. in (12.46), that

$$-\int_{\Omega\times]0,T[}(T-t)\,\Delta\,\varphi(h\cdot\nabla\varphi)dxdt = -\int_{\Gamma\times]0,T[}(T-t)\left|\frac{\partial\varphi}{\partial\nu}\right|^2 d\Gamma dt$$

$$+\int_{\Omega\times]0,T[}(T-t)\nabla\varphi\cdot\nabla(h\cdot\nabla\varphi)dxdt$$

$$=-\int_{\Gamma\times]0,T[}(T-t)\left|\frac{\partial\varphi}{\partial\nu}\right|^2 d\Gamma dt \qquad (12.49)$$

$$+\int_{\Omega\times]0,T[}(T-t)(\nabla\varphi\cdot\nabla h_k)\frac{\partial\varphi}{\partial x_k}dxdt$$

$$+\int_{\Omega\times]0,T[}(T-t)\frac{1}{2}\left(h\cdot\nabla(|\nabla\varphi|^2)\right)dxdt$$

$$=-\int_{\Gamma\times]0,T[}(T-t)\left|\frac{\partial\varphi}{\partial\nu}\right|^2 d\Gamma dt$$

$$+O(\|\psi\|^2)+\int_{\Omega\times]0,T[}\frac{T-t}{2}\left(h\cdot\nabla(|\nabla\varphi|^2)\right)dxdt.$$

From (12.44) we obtain

$$-\int_{\Omega\times]0,T[}(T-t)\,\Delta\,\varphi(h\cdot\nabla\varphi)dxdt = -\int_{\Gamma\times]0,T[}(T-t)\left|\frac{\partial\varphi}{\partial\nu}\right|^2 d\Gamma dt$$

$$+O(\|\psi\|^2)+\int_{\Gamma\times]0,T[}\frac{T-t}{2}|\nabla\varphi|^2 d\Gamma dt \qquad (12.50)$$

$$-\int_{\Omega\times]0,T[}\frac{T-t}{2}|\nabla\varphi|^2 div\,(h)\,dxdt.$$

Since $\varphi = 0$ on $\Gamma\times]0,T[$ we have that $\nabla\varphi$ is parallel to ν on $\Gamma\times]0,T[$.

Hence,

$$\frac{\partial \varphi}{\partial \nu} = |\nabla \varphi| \quad \text{on } \Gamma \times]0, T[. \tag{12.51}$$

Using this in (12.50) together with (12.45), we obtain

$$- \int_{\Omega \times]0,T[} (T-t)\Delta\varphi(h \cdot \nabla\varphi)dxdt = O(\|\psi\|^2) - \int_{\Gamma \times]0,T[} \frac{(T-t)}{2} \left| \frac{\partial \varphi}{\partial \nu} \right|^2 d\Gamma dt. \tag{12.52}$$

(12.48) and (12.52) together with (12.46) give us the result, namely

$$\int_{\Gamma \times]0,T[} \frac{(T-t)}{2} \left| \frac{\partial \varphi}{\partial \nu} \right|^2 d\Gamma dt = O(\|\psi\|^2) - \int_{\Omega \times]0,T[} (T-t)\psi(h \cdot \nabla\varphi)dxdt$$
$$= O(\|\psi\|^2), \tag{12.53}$$

where we have used the continuous linear mapping (12.45).
The proof of the lemma is completed.   ∎

LEMMA 12.6
  Given the problem (12.40) and the hypothesis (12.32) of Theorem 3, i.e.,

$$\begin{cases} F \in L^1(0,T; H^{-1}(\Omega)), \\ \Phi^0 \in L^2(\Omega), \ \Phi^1 \in H^{-1}(\Omega), \\ g \in L^2(\Gamma \times]0, T[). \end{cases} \tag{12.54}$$

Then the unique solution Φ of (12.40) satisfies

$$\begin{cases} \Phi \in L^\infty(0,T; L^2(\Omega)), \\ \Phi' \in L^\infty(0,T; H^{-1}(\Omega)). \end{cases}$$

PROOF   STEP 1
Let $F = 0$ and take all data smooth. Then let Φ be the solution of problem (12.40). In order to use transposition later on, we define φ as the solution of the backward problem

$$\begin{cases} \varphi'' - \Delta\varphi = \psi \in L^1(0,T; L^2(\Omega)) \text{ in } \Omega \times]0, T[, \\ \varphi(x,T) = \varphi'(x,T) = 0 \text{ in } \Omega, \\ \varphi = 0 \text{ on } \Gamma \times]0, T[. \end{cases} \tag{12.55}$$

By transposition (cf. Section 11.9), we obtain

$$\int_{\Omega \times]0,T[} F\varphi \, dx \, dt + \int_{\Gamma \times]0,T[} \frac{\partial \Phi^0}{\partial \nu} \varphi \, d\Gamma \, dt = \int_{\Omega \times]0,T[} \Phi \psi \, dx \, dt$$
$$+ \int_{\Gamma \times]0,T[} g \frac{\partial \varphi}{\partial \nu} \, d\Gamma \, dt$$
$$- \big(\Phi^1, \varphi(0)\big) + \big(\Phi^0, \varphi'(0)\big) .$$

Hence,

$$\int_{\Omega \times]0,T[} \Phi(\varphi'' - \Delta\varphi) = - \int_{\Gamma \times]0,T[} g \frac{\partial \varphi}{\partial \nu} \, d\Gamma \, dt + \big(\Phi^1, \varphi(0)\big) - \big(\Phi^0, \varphi'(0)\big) .$$
$$(12.56)$$

From Lemma 12.5, the right-hand side is a continuous linear functional in φ. Then the mapping

$$\psi \to - \int_{\Gamma \times]0,T[} g \frac{\partial \varphi}{\partial \nu} \, d\Gamma \, dt + \big(\Phi^1, \varphi(0)\big) - \big(\Phi^0, \varphi'(0)\big) \qquad (12.57)$$

is continuous on $L^1(0,T;L^2(\Omega))$. Hence, by transposition, there exists a unique weak solution Φ which belongs to the dual of $L^1(0,T;L^2(\Omega))$, i.e., to $L^\infty(0,T;L^2(\Omega))$.

STEP 2

Since $F = 0$ and $\Phi \in L^\infty(0,T;L^2(\Omega))$, it follows that

$$\Phi'' = \Delta\Phi \in L^\infty(0,T;H^{-2}(\Omega)). \qquad (12.58)$$

Hence, by the Continuous Derivatives Theorem we have

$$\Phi' \in L^\infty(0,T;H^{-1}(\Omega)). \qquad (12.59)$$

Then the lemma is proved for the case $F = 0$.

STEP 3

We have the same result for $F \neq 0$ and $F \in L^\infty(0,T;L^2(\Omega))$. We only have to add the solution Ψ of the problem below to Φ.

$$\begin{cases} \Psi'' - \Delta\Psi = F \text{ in } \Omega \times]0,T[, \\ \Psi(x,0) = \Psi'(x,0) = 0 \text{ in } \Omega, \\ \Psi = 0 \text{ on } \Gamma \times]0,T[, \end{cases} \qquad (12.60)$$

which exists and is unique in a weak sense according to Theorem 11.8, Section 11.8, Chapter 11. The proof is completed. ∎

LEMMA 12.7

We consider the problem (12.40), now with $F = 0$, and with the following hypotheses:

$$\begin{cases} g, g' \in L^2(\Gamma \times]0, T[) \\ \Phi^0 \in H^1(\Omega), \Phi^1 \in L^2(\Omega), \end{cases} \tag{12.61}$$

and the compatibility condition

$$g|_{t=0} = \Phi^0|_\Gamma. \tag{12.62}$$

Then

$$\begin{cases} \Phi, \Phi' \in L^\infty(0, T; L^2(\Omega)), \\ \Phi'' \in L^\infty(0, T; H^{-1}(\Omega)). \end{cases} \tag{12.63}$$

PROOF   STEP 1

Let Φ be the solution of (12.40) with $F = 0$. We are obliged to first assume that the data is smooth and then extend by continuity. This is done to assure the "legality" of the calculations. We introduce

$$\frac{\partial \Phi}{\partial t} = \Phi_1.$$

It is readily verified by a formal differentiation of (12.40) that Φ_1 satisfies

$$\begin{cases} \Phi_1'' - \Delta \Phi_1 = 0 \text{ in } \Omega \times]0, T[, \\ \Phi_1(x, 0) = \Phi^1, \quad \Phi_1'(x) = \Delta \Phi^0 \text{ in } \Omega, \\ \Phi_1 = g' \text{ on } \Gamma \times]0, T[. \end{cases} \tag{12.64}$$

(Note that we have implicitly assumed that the compatibility conditions on the initial data are satisfied, i.e., $\Delta \Phi(0) = \Phi''(0)$.)

It is shown in STEP 2 that the compatibility conditions stated in (12.62) ensure the uniqueness of the solution Φ to (12.40). Once the uniqueness of the solution Φ is ensured we can extend by continuity. Thus, by applying Lemma 12.6 to (12.64) we have

$$\begin{cases} \Phi, \Phi' \in L^\infty(0, T, L^2(\Omega)), \\ \Phi'' \in L^\infty(0, T, H^{-1}(\Omega)). \end{cases} \tag{12.65}$$

STEP 2

We have solution Φ_1 of the system (12.64). We are then obliged to prove the fact that the solution Φ obtained by integrating Φ_1 actually does satisfy problem (12.40), and that the solution is uniquely determined. This is where the compatibility conditions play an important role.

Let us write up an auxiliary system.

$$\begin{cases} v'' - \Delta v = 0 \text{ in } \Omega \times]0, T[, \\ v(0) = \Phi^1, \quad v'(0) = \Delta \Phi^0, \\ v|_\Gamma = g'. \end{cases} \tag{12.66}$$

Since $v(0)$, $v'(0)$, and g' lie in $L^2(\Omega)$, $H^{-1}(\Omega)$, and $L^2(\Gamma \times]0, T[)$, respectively, an application of Lemma 12.6 gives

$$\begin{cases} v \in L^\infty(0, T; L^2(\Omega)), \\ v' \in L^\infty(0, T; L^2(\Omega)), \\ v'' \in L^2(0, T; H^{-1}(\Omega)). \end{cases} \tag{12.67}$$

We now define an auxiliary function

$$w = \Phi(0) + \int_0^t v(s) ds. \tag{12.68}$$

Recall that the trace operator $T : H^k(\Omega) \to H^{k-\frac{1}{2}}(\Gamma)$ is a bounded operator. It is well known how a bounded operator can be extended to a closed one. So, taking the trace on both sides of (12.68), we obtain

$$Tw = T\left(\Phi^0 + \int_0^t v(s) ds\right) \tag{12.69}$$

$$= \Phi^0|_\Gamma + \int_0^t T(v(s)) ds. \tag{12.70}$$

Hence,

$$w|_\Gamma = \Phi^0|_\Gamma + \int_0^t g'(s) ds \tag{12.71}$$

$$= \Phi^0|_\Gamma + g(t) - g(0). \tag{12.72}$$

From (12.67) we have

$$\begin{cases} w \in C([0, T]; L^2(\Omega)) \subset L^\infty(0, T; L^2(\Omega)), \\ w' = v \in L^\infty(0, T; L^2(\Omega)), \\ w'' = v' \in L^\infty(0, T; L^2(\Omega)) \subset L^\infty(0, T; H^{-1}(\Omega)). \end{cases} \tag{12.73}$$

Then from (12.68) we have

$$w(0) = \Phi(0) = \Phi^0, \quad w'(0) = v(0) = \Phi^1 \tag{12.74}$$

and

$$w|_\Gamma = \Phi_\Gamma \text{ i.e. } \Phi^0|_\Gamma = g|_{t=0}. \tag{12.75}$$

Integrating (12.66) and using the initial conditions

$$w'' = v' = v'(0) + \int_0^t \Delta v(s)ds \tag{12.76}$$

$$= \Delta\Phi(0) + \int_0^t \Delta v(s)ds.$$

Partial integration gives

$$\int_0^t \Delta v(s)ds = \int_0^t \Delta w'(s)ds = \Delta w - \Delta w(0), \tag{12.77}$$

i.e.,

$$w'' = v' = \Delta w. \tag{12.78}$$

Hence,

$$w'' - \Delta w = 0. \tag{12.79}$$

Now, in order to show the uniqueness of Φ, we subtract (12.40) (with $F = 0$) from (12.78) and get

$$(w - \Phi)'' - \Delta(w - \Phi) = 0. \tag{12.80}$$

From (12.74) we have the following equalities for the initial data

$$(w - \Phi)(0) = 0, \quad (w - \Phi)'(0) = 0. \tag{12.81}$$

On the boundary, we have, due to the compatibility conditions,

$$(w - \Phi)|_\Gamma = \Phi^0 + g(t) - g|_{t=0} - g(t) = \Phi^0 - g|_{t=0} = 0. \tag{12.82}$$

We have, therefore, the following system

$$\begin{cases} (w - \Phi)'' - \Delta(w - \Phi) = 0 \text{ in } \Omega \times]0, T[, \\ (w - \Phi)(0) = (w - \Phi)'(0) = 0 \text{ in } \Omega, \\ (w - \Phi) = 0 \text{ on } \Gamma \times]0, T[. \end{cases} \tag{12.83}$$

We have, by applying the uniqueness of the solution in Lemma 12.6 to (12.83), that

$$w = \Phi \text{ in } [0, T] \text{ or } v = w' = \Phi'. \tag{12.84}$$

The proof is thus complete.   ∎

LEMMA 12.8

We consider once again the problem (12.40) with $F = 0$. We assume (12.61) and (12.62), i.e.,

$$\begin{cases} g, g' \in L^2(\Gamma \times]0, T[) \\ \Phi^0 \in H^1(\Omega), \Phi^1 \in L^2(\Omega), \end{cases} \tag{12.85}$$

and

$$g|_{t=0} = \Phi^0|_\Gamma, \tag{12.86}$$

and moreover,

$$g \in L^\infty(0, T; H^{\frac{1}{2}}(\Gamma)). \tag{12.87}$$

In addition to (12.63) we have

$$\Phi \in L^\infty(0, T; H^1(\Omega)). \tag{12.88}$$

PROOF

In this proof we shall use some classical regularity results for the Dirichlet problem for elliptic operators. Since $F = 0$, we have according to (12.63)

$$\Delta\Phi = \Phi'' \in L^\infty(0, T; H^{-1}(\Omega)). \tag{12.89}$$

By (12.87) we have

$$\Phi|_\Gamma = g \in L^\infty(0, T; H^{\frac{1}{2}}(\Gamma)). \tag{12.90}$$

Then, by assuming t to be a parameter, we have a problem of the following type

$$\begin{cases} \Delta u = f \in \Omega, \text{ where } f \in H^{-1}(\Omega), \\ u|_\Gamma = g \text{ where } g \in H^{\frac{1}{2}}(\Gamma), \end{cases}$$

i.e., a Dirichlet problem for an elliptic system. Classical regularity results for elliptic boundary value problems imply that $u \in H^1(\Omega)$. (See Agmon, Douglas, and Nirenberg [2] for a presentation of the classical regularity results for elliptic boundary value problems.) Hence,

$$\Phi \in L^\infty(0, T; H^1(\Omega)), \tag{12.91}$$

and the proof is complete.   ∎

REMARK 12.5   As mentioned in Section 12.1.3, a regularity for g weaker than $H^k(\Gamma \times]0, T[)$ is sufficient to obtain the interior regularity. The full strength of the assumption $g \in H^k(\Gamma \times]0, T[)$ is used to obtain a regularity result for $\frac{\partial \Phi}{\partial \nu}$. The latter will be made clear in STEP 2 of the proof of Theorem 1.   ∎

12.2.1   Proofs of Theorems from 12.1.2

We are now ready to prove the theorems stated in Section 12.1.2. (In order to facilitate the reading of the proofs, the theorems of Section 12.1.2 have been restated before each proof.) The problem to be solved is (12.40).

THEOREM 12.9
Consider problem (12.40). Hypotheses:

$$\begin{cases} F \in L^1(0, T; L^2(\Omega)), \\ \Phi^0 \in H^1(\Omega), \ \Phi^1 \in L^2(\Omega), \\ g \in H^1(\Gamma \times]0, T[) \end{cases} \tag{12.92}$$

with the compatibility conditions

$$g|_{t=0} = \Phi^0|_\Gamma. \tag{12.93}$$

Then the unique solution Φ of (12.40) satisfies

$$\begin{cases} \Phi \in C([0, T]; H^1(\Omega)), \\ \Phi' \in C([0, T]; L^2(\Omega)) \end{cases} \tag{12.94}$$

and

$$\frac{\partial \Phi}{\partial \nu} \in L^2(\Gamma \times]0, T[). \tag{12.95}$$

PROOF   We carry out the proof in two steps.
STEP 1 *Proof of (12.94)*
If we consider Ψ defined as in (12.60), it satisfies the property (12.94). Therefore, it suffices to consider the case $F = 0$. Since $g \in H^1(\Gamma \times]0, T[)$, we have

$$g \in L^2(0, T; H^1(\Gamma)), \tag{12.96}$$
$$g' \in L^2(0, T; L^2(\Gamma)). \tag{12.97}$$

From the Continuous Derivatives Theorem, with $m = 1$, $X = H^1(\Gamma)$ and $Y = L^2(\Gamma)$ we have $g \in C_b([0, T]; [H^1(\Gamma), H^0(\Gamma)]_{\frac{1}{2}}$. Hence,

$$g \in L^\infty(0, T; H^{\frac{1}{2}}(\Gamma)). \tag{12.98}$$

We now have the identical situation as that of Lemma 12.8. Hence, (12.94) follows from (12.63) and (12.88).

STEP 2

This step resembles the proof of STEP 2 of Lemma 12.5. The calculations are long and tedious. We have therefore only stated the important intermediate results in the calculations. We prove now that $\frac{\partial \Phi}{\partial \nu} \in L^2(\Gamma \times]0, T[)$. We assume, as usual, all data to be smooth to guarantee the validity of the calculations. As in the proof of Lemma 12.5, we have only to prove that

$$\left\| \sqrt{(T-t)} \frac{\partial \Phi}{\partial \nu} \right\|_{L^2(\Gamma \times]0, T[)} \leq CE, \tag{12.99}$$

where

$$E = \|F\|_{L^1(0,T;L^2(\Omega))} + \|\Phi^0\|_{H^1(\Omega)} + \|\Phi^1\|_{L^2(\Omega)} + \|g\|_{H^1(\Gamma \times]0,T[)}. \tag{12.100}$$

We already know from STEP 1 that

$$\| \, | \, \nabla \Phi \, | \, \|_{L^\infty(0,T;L^2(\Omega))} + \left\| \frac{\partial \varphi}{\partial t} \right\|_{L^\infty(0,T;L^2(\Omega))} \leq CE \tag{12.101}$$

where C is a constant. We now multiply (12.40) by $(T-t)h \cdot \nabla \Phi$ as in the proof of Lemma 12.5 . Collecting the steps (12.48) and (12.49) (and replacing φ and ψ by Φ and F, respectively), we have

$$\int_{\Omega \times]0,T[} (T-t) \left[\Phi'' - \Delta \Phi h \cdot \nabla \Phi dx dt = \int_{\Omega \times]0,T[} (T-t)F(h \cdot \nabla \Phi) dx dt \right.$$

$$= -T(\Phi^1, h \cdot \nabla \Phi(0))_\Omega + \int_{\Omega \times]0,T[} \Phi'(h \cdot \nabla \Phi) dx dt$$

$$- \int_{\Gamma \times]0,T[} \frac{T-t}{2} (g')^2 d\Gamma dt$$

$$- \int_{\Gamma \times]0,T[} (T-t) \left| \frac{\partial \Phi}{\partial \nu} \right| d\Gamma dt$$

$$+ \int_{\Omega \times]0,T[} \frac{T-t}{2} (\Phi')^2 (div \, h) dx dt$$

$$+ \int_{\Omega \times]0,T[} (T-t) \nabla \Phi \cdot \nabla(h \cdot \nabla \Phi) dx dt.$$

We deduce that

$$\int_{\Omega \times]0,T[} (T-t) \nabla \Phi \cdot \nabla(h \cdot \nabla \Phi) dx dt = \int_{\Gamma \times]0,T[} \frac{T-t}{2} | \nabla \Phi |^2 \, d\Gamma dt + O(F^2)$$

$$- \int_{\Omega \times]0,T[} \frac{T-t}{2} | \nabla \Phi |^2 (div \, h) dx dt. \tag{12.102}$$

Using (12.102), rearranging the terms and using the notation $O(E)$ for any quantity bounded in absolute value by CE, we obtain

$$-\int_{\Gamma\times]0,T[} (T- t\ )\left|\frac{\partial\Phi}{\partial\nu}\right|^2 d\Gamma dt + \int_{\Gamma\times]0,T[} \frac{(T-t)}{2}\mid\nabla\Phi\mid^2 d\Gamma dt$$

$$= \int_{\Gamma\times]0,T[} \frac{T-t}{2}\mid\nabla\Phi\mid^2 (div\ h)dxdt$$

$$+ \int_{\Omega\times]0,T[} (T-t)F(h\cdot\nabla\Phi)dxdt + T(\Phi^1, h\cdot\nabla\Phi^0)$$

$$- \int_{\Omega\times]0,T[} \Phi'(h\cdot\nabla\Phi)dxdt + \int_{\Gamma\times]0,T[} \frac{T-t}{2}(g')^2 d\Gamma dt$$

$$- \int_{\Omega\times]0,T[} \frac{T-t}{2}(\Phi')^2 (div\ h)dxdt + O(\|F\|^2)$$

$$= O(E^2).$$

Hence,

$$\int_{\Gamma\times]0,T[} \frac{T-t}{2}\mid\nabla\Phi\mid^2 d\Gamma dt - \int_{\Gamma\times]0,T[} (T-t)\left|\frac{\partial\Phi}{\partial\nu}\right|^2 = O(E^2). \quad (12.103)$$

Observing that

$$\frac{\partial\Phi}{\partial x_j} = \nu_j\frac{\partial\Phi}{\partial\nu} + T_j g \quad (12.104)$$

on Γ, T_j is a first-order operator on Γ with smooth coefficients (with Γ smooth), we obtain first

$$\int_{\Gamma\times]0,T[} \frac{T-t}{2}\mid\nabla\Phi\mid^2 d\Gamma dt = \int_{\Gamma\times]0,T[} \frac{T-t}{2}\left|\frac{\partial\Phi}{\partial\nu}\right|^2 d\Gamma dt + O(\|g\|^2_{L^2(0,T;H^1(\Gamma))}),$$
$$(12.105)$$

and finally (12.103) and (12.105) give

$$\int_{\Gamma\times]0,T[} \frac{T-t}{2}\left|\frac{\partial\Phi}{\partial\nu}\right|^2 d\Gamma dt = O(E^2). \quad (12.106)$$

The proof of the theorem is now complete.   ∎

REMARK 12.6   Note that in the proof of Theorem 12.9 we have actually achieved an additional regularity result,

$$\Phi'' \in L^\infty(0,T;H^{-1}(\Omega)). \quad (12.107)$$

This is a result of the proof of Lemma 12.7.   ∎

THEOREM 12.10
   Consider the system (12.40). Hypotheses :

$$
\begin{cases}
F \in L^1(0,T;H^1(\Omega)), \\
F' \in L^1(0,T;L^2(\Omega)), \\
\Phi^0 \in H^2(\Omega), \ \Phi^1 \in H^1(\Omega), \\
g \in H^2(\Gamma\times]0,T[)
\end{cases}
\tag{12.108}
$$

with the compatibility conditions

$$
g|_{t=0} = \Phi^0|_\Gamma, \ g'|_{t=0} = \Phi^1|_\Gamma.
\tag{12.109}
$$

Then the unique solution Φ of (12.40) satisfies

$$
\begin{cases}
\Phi \in C([0,T];H^2(\Omega)), \\
\Phi' \in C([0,T];H^1(\Omega)), \\
\Phi'' \in C([0,T];L^2(\Omega))
\end{cases}
\tag{12.110}
$$

and

$$
\frac{\partial \Phi}{\partial \nu} \in H^1(\Gamma\times]0,T[).
\tag{12.111}
$$

PROOF
   We carry out this proof in three steps.
STEP 1
We introduce

$$
\Phi_1 = \frac{\partial \Phi}{\partial t}.
\tag{12.112}
$$

It is readily verified that

$$
\begin{cases}
\Phi_1'' - \Delta\Phi_1 = F', \\
\Phi_1(0) = \Phi^1 , \ \Phi_1'(0) = \Delta\Phi^0 + F(0), \\
\Phi_1 = g' \text{ on } \Gamma\times]0,T[.
\end{cases}
\tag{12.113}
$$

We see that $\Phi_1(0) \in H^1(\Omega)$, $\Phi_1'(0) \in L^2(\Omega)$. It is also readily seen that $\Phi_1 = g' \in H^1(\Gamma\times]0,T[)$ (recall the definition of $H^s(\Gamma\times]0,T[)$. According to the compatibility conditions stated in (12.109), we have by an application of Theorem 1

$$
\begin{cases}
\Phi_1 = \Phi' \in L^\infty(0,T;H^1(\Omega)), \\
\Phi_1' = \Phi'' \in L^\infty(0,T;L^2(\Omega))
\end{cases}
\tag{12.114}
$$

and

$$\frac{\partial \Phi_1}{\partial \nu} \in L^2(\Gamma \times]0, T[), \tag{12.115}$$

i.e.,

$$\frac{\partial}{\partial t}\left(\frac{\partial \Phi}{\partial \nu}\right) \in L^2(\Gamma \times]0, T[). \tag{12.116}$$

STEP 2

$F' \in L^1(0, T, L^2(\Omega))$ means that F is an absolutely continuous function on $[0, T]$. Hence it follows for an even stronger reason from (12.108) that $F \in L^\infty(0, T; L^2(\Omega))$, so that by (12.114) we have

$$\begin{cases} \Delta \Phi = \Phi'' - F \in L^\infty(0, T; L^2(\Omega)), \\ \Phi|_\Gamma = g \in L^\infty(0, T, H^{\frac{3}{2}}(\Omega)). \end{cases} \tag{12.117}$$

The last line in (12.117) is seen by the application of the Continuous Derivatives Theorem. Since $g \in L^2(0, T; H^2(\Gamma)) \cap H^2(0, T; L^2(\Gamma))$, we have by choosing $m = 2$ in Theorem 10.5 that

$$g \in L^\infty(0, T, [H^0(\Gamma); H^2(\Gamma)]_\theta)) = L^\infty(0, T, H^{\frac{3}{2}}(\Omega)).$$

Now using t as the parameter, an application of the classical Dirichlet regularity for the elliptic boundary value problems implies that

$$\Phi \in L^\infty(0, T; H^1(\Gamma)). \tag{12.118}$$

STEP 3

It remains to be shown that

$$\frac{\partial \Phi}{\partial \nu} \in L^2(0, T, H^1(\Gamma)). \tag{12.119}$$

In order to do this, we introduce the so-called tangential operator

$$B = \sum_i b_i(x) \frac{\partial}{\partial x_i}. \tag{12.120}$$

B is a first-order operator (time independent) with coefficients b_i smooth in $\overline{\Omega}$ and such that B is tangent to Γ, i.e., $\sum_i b_i \nu_i = 0$ on Γ. Proving (12.119) is equivalent to proving that

$$\frac{\partial}{\partial \nu} B\Phi \in L^2(\Gamma \times]0, T[). \tag{12.121}$$

It is readily observed that:

$B\Delta - \Delta B = R$, where R is a second-order operator with smooth coefficients.

So if we set

$$B\Phi = w, \tag{12.122}$$

then it is verified that

$$\begin{cases} w'' - \Delta w = BF + R\Phi, \\ w(x,0) = B\Phi^0 \in H^1(\Omega), \\ w'(x,0) = B\Phi^1 \in L^2(\Omega), \\ w|_{\Gamma \times]0,T[} = Bg \in H^1(\Gamma \times]0,T[). \end{cases} \tag{12.123}$$

The last line in (12.123) requires some explanation. From (12.108) we have $g \in H^2(\Gamma \times]0,T[)$. Hence,

$$g \in L^2(0,T;H^2(\Gamma)) \cap H^2(0,T;L^2(\Gamma)), \tag{12.124}$$

i.e.,

$$\begin{cases} g \in L^2(0,T;H^2(\Gamma)), \\ g'' \in L^2(0,T;L^2(\Gamma)). \end{cases} \tag{12.125}$$

Since B is a first order operator, we have

$$\begin{cases} Bg \in L^2(0,T;H^1(\Gamma)), \\ Bg'' = (Bg)'' \in L^2(0,T;H^{-1}(\Gamma)). \end{cases} \tag{12.126}$$

An application of the Intermediate Derivatives Theorem gives

$$(Bg)' \in L^2(0,T;L^2(\Omega)). \tag{12.127}$$

Hence, we conclude that $Bg \in H^1(\Gamma \times]0,T[)$. Since B is tangential, the first compatibility condition in (12.109) implies

$$Bg|_{t=0} = B\Phi^0|_\Gamma. \tag{12.128}$$

Now, using (12.118) we have

$$R\Phi \in L^\infty(0,T;L^2(\Omega)). \tag{12.129}$$

Moreover, from (12.108)

$$BF \in L^1(0,T;L^2(\Omega)). \tag{12.130}$$

Thus,

$$BF + R\Phi \in L^1(0,T,L^2(\Omega)). \tag{12.131}$$

We can therefore conclude (12.121) by applying Theorem 12.9 to (12.123), so the proof is complete.   ■

THEOREM 12.11
Consider the system (12.40). Hypotheses:

$$\begin{cases} F \in L^1(0,T; H^{-1}(\Omega)), \\ \Phi^0 \in L^2(\Omega), \ \Phi^1 \in H^{-1}(\Omega), \\ g \in L^2(\Gamma \times]0,T[) \end{cases} \quad (12.132)$$

without any compatibility conditions this time.
Then the unique solution Φ of (12.40) satisfies

$$\begin{cases} \Phi \in C([0,T]; L^2(\Omega)), \\ \Phi' \in C([0,T]; H^{-1}(\Omega)) \end{cases} \quad (12.133)$$

and

$$\frac{\partial \Phi}{\partial \nu} \in H^{-1}(\Gamma \times]0,T[). \quad (12.134)$$

PROOF
   STEP 1
Lemma 12.6 gives the interior regularity of the solution.
STEP 2
It remains to show that

$$\frac{\partial \Phi}{\partial \nu} \in H^{-1}(\Gamma \times]0,T[). \quad (12.135)$$

Define φ as the solution of the system below.

$$\begin{cases} \varphi'' - \Delta \varphi = 0 \text{ in } \Omega \times]0,T[, \\ \varphi(x,0) = \varphi'(x,0) = 0 \text{ in } \Omega, \\ \varphi = h \text{ on } \Gamma \times]0,T[, \end{cases} \quad (12.136)$$

where h is given such that

$$h \in H^1(\Gamma \times]0,T[), \quad h(x,T) = 0 \text{ on } \Gamma$$

(the regularity of the solution φ is given by Theorem 12.9).
In problem (12.40) we take all data smooth and perform an integration by parts and obtain

$$\int_{\Gamma \times]0,T[} \frac{\partial \Phi}{\partial \nu} h \, d\Gamma \, dt = \int_{\Gamma \times]0,T[} F \varphi \, dx \, dt + \int_{\Gamma \times]0,T[} g \frac{\partial \varphi}{\partial \nu} \, d\Gamma \, dt \quad (12.137)$$
$$- \left(\Phi^1, \varphi(0) \right) + \left(\Phi^0, \varphi'(0) \right). \quad (12.138)$$

Using Remark 12.2 on the problem 12.136, we deduce that the right-hand side of (12.137) is a continuous linear functional of h. Hence,

$$\left| \int_{\Gamma \times]0,T[} \left(\frac{\partial \Phi}{\partial \nu} \right) h \, d\Gamma \, dt \right| \leq c \|h\|_{H^1(\Gamma \times]0,T[)}. \qquad (12.139)$$

If we choose $h \in H_0^1(\Gamma \times]0, T[)$ (i.e., $h(\cdot, 0) = 0$ and $h(\cdot, T) = 0$ on Γ), we immediately obtain the result and the proof is completed. ∎

12.3   Interpolation Results

Interpolation between Theorem 12.9 and Theorem 12.10 gives the following theorem.

THEOREM 12.12

Consider the problem (12.40). Hypotheses:

$$\begin{cases} F \in L^1(0,T; H^\theta(\Omega)), D^\theta F \in L^1(0,T; L^2(\Omega)), 0 \leq \theta \leq 1, \theta \neq \frac{1}{2} \\ \Phi^0 \in H^{1+\theta}(\Omega), \ \Phi^1 \in H^\theta(\Omega), \\ g \in H^{1+\theta}(\Gamma \times]0,T[) \end{cases}$$

$$(12.140)$$

with the compatibility conditions

$$g|_{t=0} = \Phi^0|_\Gamma, \quad \frac{\partial g}{\partial t}\Big|_{t=0} = \Phi^1|_\Gamma, \text{ if } \theta > \frac{1}{2}. \qquad (12.141)$$

Then the unique solution Φ of (12.40) satisfies

$$\begin{cases} \Phi \in C([0,T]; H^{1+\theta}(\Omega)), \\ \Phi' \in C([0,T]; H^\theta(\Omega)), \\ \Phi'' \in C([0,T]; H^{\theta-1}(\Omega)) \end{cases} \qquad (12.142)$$

and

$$\frac{\partial \Phi}{\partial \nu} \in H^\theta(\Gamma \times]0,T[). \qquad (12.143)$$

The result is simply obtained by applying the rules for interpolation between the respective spaces for the data F, Φ^0, Φ^1, and the regularity

results stated in Theorems 1 and 2, respectively.
According to Theorems 1 and 2, the mapping (c.f. Remark 12.2)

$$\{F, \Phi^0, \Phi^1, g\} \rightarrow \left\{\Phi, \ \Phi', \ \frac{\partial \Phi}{\partial \nu}\right\} \tag{12.144}$$

is a continuous linear mapping of

$$L^1(0, T, L^2(\Omega)) \times H^1(\Omega) \times L^2(\Omega) \times H^1(\Omega) \text{ into}$$
$$C([0, T]; H^1(\Omega)) \times C([0, T]; L^2(\Omega)) \times L^2(\Gamma \times]0, T[)$$

and

$$L^1(0, T, L^2(\Omega)) \times H^2(\Omega) \times H^1(\Omega) \times H^2(\Omega) \text{ into}$$
$$C([0, T]; H^2(\Omega)) \times C([0, T]; H^1(\Omega)) \times H^1(\Gamma \times]0, T[).$$

If we consider these two mappings as a couple of continuous linear mappings denoted by π, a simple application of the Main Theorem of Interpolation, Theorem 10.7, gives us the following interpolation results.

Let us first consider the data.

$$F \in [L^1(0, T, H^1(\Omega)), L^1(0, T, H^0(\Omega))]_\theta = L^1(0, T; [H^1(\Omega), H^0(\Omega)]_\theta)$$
$$= L^1(0, T, H^{1-\theta}(\Omega)). \quad (12.145)$$

We also demand that $D^\theta F \in L^1(0, T; L^2(\Omega)), 0 \le \theta \le 1, \theta \ne \frac{1}{2}$ (in order to obtain Theorem 2). Similarly, for Φ^0, Φ^1, and g we have

$$\Phi^0 \in [H^2(\Gamma \times]0, T[), H^1(\Gamma \times]0, T[)]_\theta = H^{2-\theta}(\Omega)$$
$$\Phi^1 \in [H^1(\Gamma \times]0, T[), H^0(\Gamma \times]0, T[)]_\theta = H^{1-\theta}(\Omega)$$
$$g \in [H^2(\Gamma \times]0, T[), H^1(\Gamma \times]0, T[)]_\theta = H^{2-\theta}(\Omega).$$

The interpolation spaces in which the solution (to Problem (12.40)) and its derivatives $\{\Phi, \Phi', \Phi''\}$ lie are given by

$$\Phi \in L^\infty(0, T, [H^2(\Omega), H^1(\Omega)]_\theta) \equiv L^\infty(0, T, H^{2-\theta}(\Omega))$$
$$\Phi' \in L^\infty(0, T, [H^1(\Omega), H^0(\Omega)]_\theta) \equiv L^\infty(0, T, H^{1-\theta}(\Omega))$$
$$\Phi'' \in L^\infty(0, T, [H^0(\Omega), H^{-1}(\Omega)]_\theta) \equiv L^\infty(0, T, H^{-\theta}(\Omega))$$

Now, setting $\tilde{\theta} = 1 - \theta$, we have exactly the hypotheses stated in (12.140) and (12.141), thereby obtaining the regularity results in (12.142). ∎

12.4   Some Additional Regularity Theorems

Under stronger hypothesis on the data, it is possible to achieve stronger regularity theorems.

THEOREM 12.13

We consider once again the system (12.40). Hypotheses :

$$\begin{cases} F \in L^1(0,T;H^2(\Omega)), \; F' \in L^1(0,T;H^1(\Omega)), \; F'' \in L^1(0,T;L^2(\Omega)), \\ \Phi^0 \in H^3(\Omega), \Phi^1 \in H^2(\Omega) \\ g \in H^3(\Omega) \end{cases}$$

$$(12.146)$$

with the compatibility conditions (12.93) and

$$g''|_{t=0} = \Delta\Phi^0 + F(0) \text{ on } \Gamma. \qquad (12.147)$$

Then

$$\begin{cases} \Phi \in C([0,T];H^3(\Omega)) \\ \Phi' \in C([0,T];H^2(\Omega)) \\ \Phi'' \in C([0,T];H^1(\Omega)) \\ \Phi''' \in C([0,T];L^2(\Omega)) \end{cases}$$

$$(12.148)$$

and

$$\frac{\partial\Phi}{\partial\nu} \in H^2(\Gamma\times]0,T[). \qquad (12.149)$$

PROOF
See Lasiecka, Lions, and Triggiani [16], pp. 165-166.   ∎

REMARK 12.7   By comparing Theorems 12.10 and 12.13, we see that there is an increase by one unit in the regularity of the derivatives in all directions both in space and time. It is therefore not surprising that we can proceed to any degree of regularity.   ∎

Remark 12.7 above is made explicit in the following theorem.

THEOREM 12.14

Hypotheses :

$$\begin{cases} F \in L^1(0,T;H^m(\Omega)), & F^m \in L^1(0,T;L^2(\Omega)), \\ \Phi^0 \in H^{m+1}(\Omega), & \Phi^1(\Omega) \in H^m(\Omega), \\ g \in H^{m+1}(\Gamma \times]0,T[). \end{cases} \tag{12.150}$$

We also assume that all the compatibility conditions which make sense are fulfilled. Then

$$\begin{cases} \Phi \in C([0,T];H^{m+1}(\Omega)), \\ \Phi^{m+1} \in C([0,T];L^2(\Omega)) \end{cases} \tag{12.151}$$

and

$$\frac{\partial \Phi}{\partial \nu} \in H^m(\Omega). \tag{12.152}$$

12.5 Systems with Variable Coefficients

We have until now proved a number of regularity theorems for the classical wave equation with Dirichlet boundary conditions. In physical applications, the media in which the waves propagate are often nonhomogeneous, and systems that describe the evolution in such media contain operators with variable coefficients. The study of solvability theorems for systems with variable coefficients, which is carried out in this section, is therefore highly relevant.

We are actually able to (after some modifications) reproduce all the theorems from Section 12.2. As an illustration, we prove in this section the corresponding generalization of Theorem 1 (from which Theorems 2 and 3 are derived).

12.5.1 A Regularity Theorem

THEOREM 12.15

We consider the problem

$$\begin{cases} \Phi'' + A(x,t)\Phi = F & \text{in } \Omega \times]0,T[, \\ \Phi(x,0) = \Phi^0, \ \Phi'(x,0) = \Phi^1 & \text{in } \Omega, \\ \Phi = g & \text{on } \Gamma \times]0,T[, \end{cases} \tag{12.153}$$

where

$$A = -\frac{\partial}{\partial x_i}\left(a_{ij}(x,t)\frac{\partial}{\partial x_j}\right),$$

$$a_{ij} = a_{ji}, \quad a_{ij}(x,t), \quad a'_{ij}(x,t) \in L^\infty(\Omega\times]0,T[),$$

(12.154)

$$\frac{\partial}{\partial x_k}(a_{ij}(x,t)) \in L^\infty(\Omega\times]0,T[),$$

$$a_{ij}(x,t)\xi_i\xi_j > \alpha\xi_i\xi_i \ \forall\xi_i \in R, \ \alpha > 0, \ \text{a.e. in } \Omega\times]0,T[.$$

Hypotheses:

$$\begin{cases} F \in L^1(0,T;L^2(\Omega)), \\ \Phi^0 \in H^1(\Omega), \\ \Phi^1 \in L^2(\Omega), \\ g \in H^1(\Gamma\times]0,T[) \end{cases}$$

(12.155)

with the compatibility condition

$$g|_{t=0} = \Phi^0|_\Gamma.$$

(12.156)

Then we have a solution Φ *satisfying*

$$\|\Phi\|_{C([0,T];H^1(\Omega))} + \|\Phi'\|_{C([0,T];L^2(\Omega))} + \|\frac{\partial\Phi}{\partial\nu}\|_{L^2(\Omega)}$$
$$\leq C\left\{\|F\|_{L^1(0,T;L^2(\Omega))} + \|\Phi^0\|_{H^1(\Omega)} + \|\Phi^1\|_{L^2(\Omega)} + \|g\|_{H^1(\Omega)}\right\}. \quad (12.157)$$

PROOF
In this proof we introduce the notation:

$$\begin{cases} a(t;\varphi,\psi) = \int_\Omega a_{ij}(x,t)\frac{\partial\varphi}{\partial x_j}\frac{\partial\psi}{\partial x_i}dx \\ a'(t;\varphi,\psi) = \int_\Omega a'_{ij}(x,t)\frac{\partial\varphi}{\partial x_j}\frac{\partial\psi}{\partial x_i}dx \\ a(t,\varphi) = a(t;\varphi,\varphi) \\ a'(t,\varphi) = a'(t;\varphi,\varphi) \end{cases}$$

(12.158)

and the co-normal derivative with respect to the operator A

$$\frac{\partial\varphi}{\partial\nu_A} = a_{ij}(x,t)\nu_i\frac{\partial\varphi}{\partial x_j}.$$

(12.159)

STEP 1
We show (12.157) with all data being smooth and with $L^\infty(]0,T[)$ instead
of $C([0,T])$. Hereafter, we extend by continuity using the compatibility

condition.

If we multiply (12.153) by Φ' and integrate over Ω, we obtain

$$\frac{1}{2}\frac{\partial\|\Phi'(t)\|^2_{L^2(\Omega)}}{\partial t} - \int_\Gamma \frac{\partial\Phi}{\partial\nu_A}\varphi' d\Gamma + a(t; \Phi(t), \Phi'(t)) = \int_\Omega F\Phi' dx. \quad (12.160)$$

Since A is symmetric, we note that

$$\frac{d}{dt}a(t; \Phi(t), \Phi(t)) = a'(t; \Phi(t), \Phi(t)) + 2a(t; \Phi(t), \Phi(t))$$

which, in combination with (12.160), gives

$$\frac{d}{dt}[\|\Phi'(t)\|^2_{L^2(\Omega)} + a(t; \Phi(t))] = a'(t; \Phi(t)) + 2\int_\Gamma \frac{\partial\Phi}{\partial\nu_A}g' d\Gamma + 2\int_\Omega F\Phi' dx.$$

Hence,

$$\|\Phi'(t)\|^2_{L^2(\Omega)} + a(t; \Phi(t)) = \int_0^t a'(s; \Phi(s))ds + \|\Phi^1\|^2_{L^2(\Omega)}$$

$$+a(0; \Phi^0) + 2\int_0^t \int_\Omega F\Phi' dx ds$$

$$+2\int_0^t \int_\Gamma \frac{\partial\Phi}{\partial\nu_A}g' d\Gamma dt. \quad (12.161)$$

Using the $L^\infty(\Omega\times]0, T[)$ regularity of $a_{ij}(x, t)$ and $a'_{ij}(x, t)$ on the R.H.S. of (12.161) and the ellipticity of A on the left-hand side, we obtain

$$\|\Phi'\|^2_{L^2(\Omega)} + \alpha\|\nabla\Phi(t)\|^2_{L^2(\Omega)} \le C\left\{\int_0^t |\nabla\Phi(s)|^2 ds + \|\Phi^1\|^2_{L^2(\Omega)}\right\}$$

$$+C\left\{\|\Phi^0\|^2_{H^1(\Omega)} + \left|\int_0^t \int_\Omega F\Phi' dx ds\right|\right\}$$

$$+2\left|\int_0^t \int_\Gamma \frac{\partial\Phi}{\partial\nu_A}g' d\Gamma dt\right|.$$

$$(12.163)$$

STEP 2

We will now estimate the normal derivative $\frac{\partial\Phi}{\partial\nu}$. The proof is similar to the proof in the case of constant coefficients (c.f. STEP 2 in Lemma 12.5 and Theorem 12.9). With the functions $h_k \in C^1(\bar\Omega)$ and $h_k = \nu_k$ on Γ, we will in this case use the multiplier

$$h_k\left(\frac{\partial\Phi}{\partial x_k}\right). \quad (12.164)$$

First, multiplying $\Phi''(t)$ by (12.164) and integrating by parts over $\Omega\times]0, t[$, we obtain

$$\int_\Omega \int_0^t \Phi'' h_k \frac{\partial \Phi}{\partial x_k} ds dx = \int_\Omega \Phi'(t) h_k \frac{\partial \Phi(t)}{\partial x_k} dx - \left(\Phi^1, h_k \frac{\partial \Phi^0}{\partial x_k}\right)$$

$$\underbrace{-\frac{1}{2}\int_\Omega \int_0^t h_k \frac{\partial (\Phi')^2}{\partial x_k} ds dx}_{I} \qquad (12.165)$$

$$= \int_\Omega \Phi'(t) h_k \frac{\partial \Phi(t)}{\partial x_k} dx - \left(\Phi^1, h_k \frac{\partial \Phi^0}{\partial x_k}\right)$$

$$-\frac{1}{2}\int_0^t \int_\Gamma (g')^2 d\Gamma dt$$

$$+\frac{1}{2}\int_0^t \int_\Omega (\Phi')^2 \frac{\partial h_k}{\partial x_k} dx dt. \qquad (12.166)$$

We have used the identity $\int_\Omega h\nabla\beta dx = \int_\Gamma \beta d\Gamma - \int_\Omega \beta div(h)dx$ in the integral I and the fact that $\Phi'|_{\Gamma\times]0,T[} = g'$.

Now, multiplication of the operator A by (12.164) and using the fact that $h_k = \nu_k$ on Γ, an integration by parts over $\Omega\times]0, t[$ gives

$$\int_0^t \int_\Omega A\Phi h_k \frac{\partial \Phi}{\partial x_k} dx dt = -\int_0^t \int_\Gamma \frac{\partial \Phi}{\partial \nu_A} \nu_k \frac{\partial \Phi}{\partial x_k} d\Gamma dt$$

$$+\int_0^t \int_\Omega a_{ij} \frac{\partial \Phi}{\partial x_j} \frac{\partial h_k}{\partial x_i} \frac{\partial \Phi}{\partial x_k} dx dt$$

$$\underbrace{+\int_0^t \int_\Omega a_{ij} \frac{\partial \Phi}{\partial x_j} h_k \frac{\partial^2 \Phi}{\partial x_i \partial x_k} dx dt}_{II}. \quad (12.167)$$

Note the identity

$$2\frac{\partial \Phi}{\partial x_j} \frac{\partial^2 \Phi}{\partial x_i \partial x_k} dx dt = \frac{\partial}{\partial x_i}\left(\frac{\partial \Phi}{\partial x_j} \frac{\partial \Phi}{\partial x_k}\right) - \frac{\partial}{\partial x_j}\left(\frac{\partial \Phi}{\partial x_k} \frac{\partial \Phi}{\partial x_i}\right) + \frac{\partial}{\partial x_k}\left(\frac{\partial \Phi}{\partial x_i} \frac{\partial \Phi}{\partial x_j}\right).$$
$$(12.168)$$

The insertion of (12.168) in the integral II and an integration by parts gives

$$II = \frac{1}{2}\int_0^t \int_\Gamma a_{ij}\nu_k \left[\nu_i \frac{\partial \Phi}{\partial x_j} \frac{\partial \Phi}{\partial x_k} - \nu_k \frac{\partial \Phi}{\partial x_i} \frac{\partial \Phi}{\partial x_j}\right] d\Gamma dt$$

$$+\frac{1}{2}\int_0^t \int_\Omega \left[-\frac{\partial}{\partial x_i}(a_{ij}h_k)\frac{\partial \Phi}{\partial x_j} \frac{\partial \Phi}{\partial x_k} + \frac{\partial}{\partial x_j}(a_{ij}h_k)\frac{\partial \Phi}{\partial x_k} \frac{\partial \Phi}{\partial x_i} - \frac{\partial}{\partial x_k}(a_{ij}h_k)\frac{\partial \Phi}{\partial x_i} \frac{\partial \Phi}{\partial x_j}\right]$$

Using Equation (12.159) and the fact that $a_{ij} = a_{ji}$ in the second term in the boundary integral above, we infer that

$$II = \frac{1}{2} \int_0^t \int_\Gamma \frac{\partial \Phi}{\partial \nu_A} \frac{\partial \Phi}{\partial \nu} - \frac{\partial \Phi}{\partial \nu_A} \frac{\partial \Phi}{\partial \nu} + a_{ij} \frac{\partial \Phi}{\partial x_i} \frac{\partial \Phi}{\partial x_j} + O\left(\int_0^t \|\nabla \Phi\|_{L^2(\Omega)}^2 ds \right),$$
(12.169)

where the assumptions on a_{ij} stated in (12.154) have been used in the last estimate above.

Now, adding (12.166) and (12.167) (where II is given by (12.169)), we obtain from (12.153)

$$\int_0^t \int_\Omega F h_k \frac{\partial \Phi}{\partial x_k} dx ds = \int_\Omega \Phi'(t) h_k \frac{\partial \Phi}{\partial x_k} dx - \left(\Phi^1, h_k \frac{\partial \Phi^0}{\partial x_k} \right) - \frac{1}{2} \int_0^t \int_\Gamma (g')^2 d\Gamma dt$$

$$+ \frac{1}{2} \int_0^t \int_\Omega (\Phi')^2 \frac{\partial h_k}{\partial x_k} dx dt - \int_0^t \int_\Gamma \frac{\partial \Phi}{\partial \nu_A} \nu_k \frac{\partial \Phi}{\partial x_k} d\Gamma dt$$

$$+ \int_0^t \int_\Gamma a_{ij} \frac{\partial \Phi}{\partial x_i} \frac{\partial \Phi}{\partial x_j} d\Gamma dt + \int_0^t \int_\Omega a_{ij} \frac{\partial \Phi}{\partial x_j} \frac{\partial \Phi}{\partial x_k} \frac{\partial h_k}{\partial x_i} dx dt$$

$$+ O\left(\int_0^t \|\nabla \Phi\|_{L^2(\Omega)}^2 ds \right).$$

Collecting the boundary terms on the L.H.S. we have

$$\underbrace{\int_0^t \int_\Gamma \frac{\partial \Phi}{\partial \nu_A} \frac{\partial \Phi}{\partial \nu} - \frac{1}{2} a_{ij} \frac{\partial \Phi}{\partial x_i} \frac{\partial \Phi}{\partial x_j} d\Gamma dt}_{III} = \qquad (12.170)$$

$$O\left(\int_0^t [\|\nabla \Phi\|_{L^2(\Omega)}^2 + \|\Phi'\|_{L^2(\Omega)}^2] ds + \int_0^t \int_\Gamma (g')^2 d\Gamma dt \right)$$

$$- \int_0^t \int_\Omega F h \cdot \nabla \Phi dx ds + \int_\Omega \Phi'(t) h \cdot \nabla \Phi^0 dx - (\Phi^1, h \cdot \nabla \Phi^0).$$

Using the co-normal derivative (12.159) and recalling that

$$\frac{\partial \Phi}{\partial \nu} = \nu_j \frac{\partial \Phi}{\partial \nu} + T_j g \quad \text{on } \Gamma, \qquad (12.171)$$

where T_j equals the first order operator on Γ, we obtain for the L.H.S. (III) of (12.170) that

$$III = \int_0^t \int_\Gamma a_{ij} \nu_i \left(\nu_j \frac{\partial \Phi}{\partial \nu} + T_i g \right) \frac{\partial \Phi}{\partial \nu} - \frac{1}{2} a_{ij} \left(\nu_i \frac{\partial \Phi}{\partial \nu} + T_i g \right) \left(\nu_j \frac{\partial \Phi}{\partial \nu} + T_j g \right) d\Gamma dt$$

$$= \int_0^t \int_\Gamma \frac{1}{2} a_{ij} \nu_i \nu_j \left(\frac{\partial \Phi}{\partial \nu} \right) + \left[a_{ij} \nu_i T_i g - \frac{1}{2} a_{ij} \nu_i T_j g - \frac{1}{2} \nu_j T_i g \right] \frac{\partial \Phi}{\partial \nu} d\Gamma dt$$

$$- \frac{1}{2} \int_0^t \int_\Gamma a_{ij} T_i g T_j g \, d\Gamma dt.$$

Since $a_{ij} = a_{ji}$, the above bracket equals zero. Hence, since A is elliptic, we obtain

$$III \geq \frac{1}{2} \int_0^t \left[\alpha \left(\frac{\partial \Phi}{\partial \nu} \right)^2 - a_{ij} T_i g T_j g \right] d\Gamma dt. \tag{12.172}$$

Hence, by (12.172) and the right-hand side of (12.170),

$$\alpha \int_0^t \int_\Gamma \left(\frac{\partial \Phi}{\partial \nu} \right)^2 d\Gamma dt \leq c \left(\int_0^t \|\nabla \Phi\|_{L^2(\Omega)}^2 + \|\Phi'\|_{L^2(\Omega)}^2 ds \right) \tag{12.173}$$

$$+ c \left(\int_0^t \int_\Gamma (g') + |\nabla_T g|^2 d\Gamma dt + \|\Phi'\|_{L^2(\Omega)} \|\nabla \Phi(t)\|_{L^2(\Omega)} \right)$$

$$+ c \left(\|\Phi^1\|_{L^2(\Omega)}^2 + \|\nabla \Phi^0\|_{L^2(\Omega)}^2 \right) + \left| \int_0^t \int_\Omega Fh \cdot \nabla \Phi dx ds \right|.$$

STEP 3

If we combine the co-normal derivative (12.159) and (12.171), we have that

$$\frac{\partial \Phi}{\partial \nu_A} = a_{ij} \nu_i \nu_j \frac{\partial \Phi}{\partial \nu} + a_{ij} \nu_i T_j g \quad \text{on } \Gamma. \tag{12.174}$$

Hence,

$$\int_0^t \int_\Gamma \left| \frac{\partial \Phi}{\partial \nu_A} \right|^2 d\Gamma dt \leq c \int_0^t \int_\Gamma \left| \frac{\partial \Phi}{\partial \nu} \right|^2 + |\nabla_T g|^2 d\Gamma dt. \tag{12.175}$$

From (12.175) we can estimate the last term in (12.162) obtained in STEP 1. For the estimate, we first apply (12.175) and then the result (12.173) of STEP 2.

$$2 \left| \int_0^t \int_\Gamma \frac{\partial \Phi}{\partial \nu_A} g' d\Gamma dt \right| \leq 2 \left[\int_0^t \int_\Gamma \left| \frac{\partial \Phi}{\partial \nu_A} \right|^2 d\Gamma dt \right]^{\frac{1}{2}} \|g'\|_{L^2(\Gamma \times]0,T[)} \tag{12.176}$$

$$\leq c \|g'\|_{L^2(\Gamma \times]0,T[)} \left[\int_0^t \int_\Gamma \left| \frac{\partial \Phi}{\partial \nu} \right|^2 + |\nabla_T g|^2 d\Gamma dt \right]^{\frac{1}{2}}$$

$$\leq c \|g'\|_{L^2(\Gamma \times]0,T[)} \{ \int_0^t \|\nabla \Phi\|_{L^2(\Omega)}^2 + \|\Phi'\|_{L^2(\Omega)}^2 ds$$

$$+ \|g\|_{H^1(\Gamma \times]0,T[)} + \|\Phi^1\|_{L^2(\Omega)} + \|\nabla \Phi^0\|_{L^2(\Omega)} + | \int_0^t \int_\Omega Fh \cdot \nabla \Phi dx ds |\}^{\frac{1}{2}}$$

$$+ c \|g'\|_{L^2(\Gamma \times]0,T[)} \left[\|\Phi'(t)\|_{L^2(\Omega)} \|\nabla \Phi(t)\|_{L^2(\Omega)} \right]^{\frac{1}{2}}.$$

If we use

$$2 \|g'\|_{L^2(\Gamma \times]0,T[)} \|\Phi'(t)\|_{L^2(\Omega)}^{\frac{1}{2}} \| \nabla \Phi(t)\|_{L^2(\Omega)}^{\frac{1}{2}} \leq \frac{1}{\epsilon} \|g'\|_{L^2(\Gamma \times]0,T[)}^2$$

$$+ \epsilon \frac{1}{2} \left(\frac{1}{\alpha} \|\Phi'(t)\|_{L^2(\Omega)}^2 + \alpha \|\nabla \Phi(t)\|_{L^2(\Omega)}^2 \right)$$

with $\epsilon < \alpha$ and $\alpha < 1$ (which can be done without affecting the ellipticity), we obtain

$$2\left|\int_0^t \int_\Gamma \frac{\partial \Phi}{\partial \nu_A} g' d\Gamma dt\right| \leq c\left\{\|g'\|^2_{L^2(\Gamma \times]0,T[)} + \int_0^t \|\nabla \Phi\|^2_{L^2(\Omega)} + \|\Phi'\|^2_{L^2(\Omega)} ds\right\}$$

$$+c\left\{\|g\|^2_{H^1(\Gamma \times]0,T[)} + \|\Phi^1\|^2_{L^2(\Omega)} + \|\Phi^0\|^2_{H^1(\Omega)}\right\} \tag{12.177}$$

$$+c\left|\int_0^t \int_\Omega Fh \cdot \nabla \Phi dx ds\right| + \frac{1}{2}\left(\|\Phi'(t)\|^2_{L^2(\Omega)} + \alpha\|\nabla \Phi(t)\|^2_{L^2(\Omega)}\right).$$

Using (12.177) in (12.162) from STEP 1, we obtain

$$\frac{1}{2}\left(\|\Phi'(t)\|^2_{L^2(\Omega)} + \alpha\|\nabla \Phi\|^2_{L^2(\Omega)} \leq c\int_0^t \|\nabla \Phi\|^2_{L^2(\Omega)} + \|\Phi'\|^2_{L^2(\Omega)} ds\right.$$

$$+ c\left\{\|g\|^2_{H^1(\Gamma \times]0,T[)} + \|\Phi^1\|^2_{L^2(\Omega)} + \|\Phi^0\|^2_{H^1(\Omega)}\right\}$$

$$+ c\int_0^t \|F(s)\|_{L^2(\Omega)} ds \sup_{0 \leq s \leq t} \|\Phi'(s)\|_{L^2(\Omega)}$$

$$+ \left(\int_0^t \|F(s)\|_{L^2(\Omega)}\right)^2. \tag{12.178}$$

If we define

$$N(t) = \|\Phi'(t)\|^2_{L^2(\Omega)} + \alpha\|\nabla \Phi\|^2_{L^2(\Omega)}$$

we can use Example 11.8 to obtain

$$\|\Phi\|_{L^\infty(0,T;H^1(\Omega))} + \|\Phi'\|_{L^\infty(0,T;L^2(\Omega))} \leq$$
$$C\left\{\|F\|_{L^1(0,T;L^2(\Omega))} + \|\Phi^0\|_{H^1(\Omega)} + \|\Phi^1\|_{L^2(\Omega)} + \|g\|_{H^1(\Omega)}\right\},$$

i.e., interior regularity. If we return to (12.173) we see that we obtain the desired bound also on $\|\frac{\partial \Phi}{\partial \nu}\|_{L^2(\Omega)}$, which together with the just-achieved interior regularity gives the result. ∎

We have now shown a number of regularity theorems for the hyperbolic mixed problem. In the next chapter we will study the question of controllability for systems of the evolution type. The theory of regularity for the hyperbolic mixed problem plays a very important role here.

Chapter 13

The Hilbert Uniqueness Method

In this chapter we consider *boundary controllability* for systems described by linear partial differential equations. So in the rest of this chapter we will assume that Ω denotes an open bounded set in R^n such that

$$\begin{cases} \text{the boundary } \Gamma \text{ of } \Omega \text{ is an } (n-1) \text{ dimensional} \\ \text{infinitely differentiable variety, } \Omega \text{ being} \\ \text{locally on one side of } \Gamma, \end{cases} \qquad (13.1)$$

unless otherwise stated. The problem that we will consider can be formulated as follows:

Let $y(x, t)$ denote the state of the system as a function of space $x \in \Omega$, and of time t. We can act on the system by "control variables". These are functions κ which are applied on Γ, the boundary of Ω, or merely on parts of the boundary. In this chapter we concentrate on linear hyperbolic partial differential equations.

The basic aim is to achieve exact controllability, i.e., *given a time T, the aim is to find for all initial data a set of controls κ that drives the system to a certain state at time T*. Since the system is assumed to be linear, this is equivalent to finding controls driving the system to rest at time T.

There are two problems involved in this, one is to find the controls κ and the other is to characterize in a suitable fashion the function spaces in which the initial data may lie in order to achieve exact controllability.

It is, however, not always possible to achieve exact controllability. We may sometimes be unable to characterize explicitly the function spaces in which the initial data lie. We are also sometimes unable to steer a system to a desired state. In this case, the system tends to the desired state in suitable norm. This is approximate controllability.

These kinds of controllability problems are highly relevant in a number of engineering disciplines, e.g., in the control of flexible space structure, plates, beams, etc. are of significant importance to engineers.

There already exists a great amount of literature that takes up this problem. In this chapter we present one of the recent methods, *The Hilbert*

Uniqueness Method (HUM) introduced by J.L. Lions in Lions [21]. The method is based on uniqueness results for the systems under consideration, and appropriate Hilbert space structures are then constructed on the space of initial data for these systems. This makes it possible to characterize the spaces in which the initial data must lie in order to guarantee exact controllability, and identify a suitable set of different controls κ. The controls found are optimal in the sense that the method minimizes the energy required to bring the system to rest. Since the method is in principle systematic, it can at be applied to a variety of problems. A variant of the method also proves to be useful in studying the question of approximate controllability.

In order to state the problem of controllability, it is necessary to ensure that the problem is well posed. This requires a number of results from PDE theory and regularity theory for hyperbolic differential operators. In fact, the improvements in the regularity results in the early eighties gave the impetus for obtaining the controllability results.

13.1   The Wave Equation - Dirichlet Boundary Control

In this section we present the general idea behind the Hilbert Uniqueness Method (HUM) by applying the method to the classical wave equation with Dirichlet boundary control. In the next section the general concept of HUM will be presented.

The problem given is to find a control κ such that for any initial conditions $\{y^0, y^1\}$ in a suitable Hilbert space we are able to drive the system

$$
\begin{cases}
y'' - \Delta y = 0, \\
y(x,0) = y^0, \quad y'(x,0) = y^1 \text{ in } \Omega, \\
y = \kappa \text{ on } \Gamma_0 \times]0, T[, \\
y = 0 \text{ on } \Gamma \backslash \Gamma_0 \times]0, T[
\end{cases}
\tag{13.2}
$$

to rest in finite time T (i.e., the solution y satisfies $y(x,T) = y'(x,T) = 0$) by using a boundary control κ on only a part Γ_0 of the boundary Γ. Since the system (13.2) is linear, this problem corresponds to drive the system (13.2) to a desired state at time T.

13.1.1   Application of HUM

The subset Γ_0 of the boundary on which we allow the control κ to act is given in Definition 13.1 below.

DEFINITION 13.1 *[Control Area] Let Ω be an open, bounded subset of R^n with smooth boundary $\partial\Omega = \Gamma$. With x^0 arbitrarily chosen in R^n, the control area Γ_0 is chosen such that*

$$\Gamma_0 = \{x \in \Gamma \mid (x - x^0)\nu \geq 0\} \tag{13.3}$$
$$= \{x \in \Gamma \mid m(x)\nu \geq 0\}, \tag{13.4}$$

where $m(x) = (x - x^0) = (x_k - x_k^0)$ and ν is the outer normal to Γ.

In other words, the scalar product between the outer normal and the vector from x^0 to x has to be greater than or equal to zero. The basic idea behind HUM consists of solving two auxiliary coupled systems. The first one is the $\varphi - system$

$$\begin{cases} \varphi'' - \Delta\varphi = 0, \\ \varphi(x,0) = \varphi^0 \,, \, \varphi'(x,0) = \varphi^1 \text{ in } \Omega, \\ \varphi = 0 \text{ on } \Gamma\times]0,T[. \end{cases} \tag{13.5}$$

By choosing $\varphi^0 \in H_0^1(\Omega)$ and $\varphi^1 \in L^2(\Omega)$, we obtain from Section 12.1.2, Chapter 12, that there exists a unique solution $\varphi \in C([0,T]; H_0^1(\Omega)\cap H^2(\Omega))$ with $\frac{\partial\varphi}{\partial\nu} \in L^2(\Gamma\times]0,T[)$. $\frac{\partial\varphi}{\partial\nu}$ is then used as a boundary condition in the second system, the ψ-system (13.6) below.

$$\begin{cases} \psi'' - \Delta\psi = 0, \\ \psi(T,x) = \psi'(T,x) = 0 \text{ in } \Omega, \\ \psi = \frac{\partial\varphi}{\partial\nu} \text{ on } \Gamma_0\times]0,T[, \\ \psi = 0 \text{ on } \Gamma \setminus \Gamma_0\times]0,T[. \end{cases} \tag{13.6}$$

The above nonhomogeneous boundary value problem admits at least a unique weak solution as stated Section 12.1.2, Chapter 12.

By using $\frac{\partial\varphi}{\partial\nu}$ as a boundary condition, the space where the solution of the $\psi - system$ lies depends naturally on the space of $\{\varphi^0, \varphi^1\}$. The solutions of both systems exist and are unique. Therefore, by construction we have uniquely defined an operator Λ:

$$\Lambda\{\varphi^0, \varphi^1\} = \{\psi'(x,0), \psi(x,0)\}. \tag{13.7}$$

If Λ is invertible, then given initial conditions $\{\psi'(x,0), \psi(x,0)\}$, we are able to find $\{\varphi^0, \varphi^1\}$, and subsequently solve the $\varphi - system$. If we then take the control κ in the y-system as

$$\kappa = \frac{\partial\varphi}{\partial\nu} \text{ on } \Gamma_0\times]0,T[, \tag{13.8}$$

and realize the connection between the y-system and the ψ-system, the problem is solved. We can drive the system (13.2) to rest in time T.

From the regularity theorems of second-order hyperbolic equations in Chapter 12, the natural space of the initial conditions $\{\psi'(x,0), \psi(x,0)\}$ will be $H^{-1}(\Omega) \times L^2(\Omega)$. The initial conditions for the system (13.2) lie in this space as stated in Theorem 13.1 below. The main problem of the method is actually to characterize the space of initial conditions, as will become evident in the forthcoming.

The above discussion is made precise in the theorem below.

THEOREM 13.1

Consider $\Gamma_0 \subset \Gamma$ defined by Definition 13.1. If

$$T > T_0 = 2R(x^0) = 2\sup_{x \in \Gamma}(x - x^0)\nu,$$

then for any initial conditions $\{y^1, y^0\} \in H^{-1}(\Omega) \times L^2(\Omega)$, one can find a control $\kappa \in L^2(\Gamma_0 \times]0, T[)$ such that κ drives the system

$$\begin{cases} y'' - \Delta y = 0, \\ y(x,0) = y^0, \ y'(x,0) = y^1 \text{ in } \Omega, \\ y = \kappa \text{ on } \Gamma_0 \times]0, T[, \\ y = 0 \text{ on } \Gamma \backslash \Gamma_0 \times]0, T[\end{cases} \qquad (13.9)$$

to rest in finite time $T = T_0$.

Some comments are in order.

REMARK 13.1   For the operator Λ defined above, we have that

Λ is an isomorphism from $H^1_0(\Omega) \times L^2(\Omega)$ onto $H^{-1}(\Omega) \times L^2(\Omega)$, (13.10)

where $H^{-1}(\Omega)$ is the dual of $H^1_0(\Omega)$. Moreover, since $\frac{\partial \varphi}{\partial \nu}$ is used as boundary condition in the ψ-system, we have that the mapping

$$g \to \{\varphi, \varphi'\} : L^2(]0, T[) \to L^2(\Omega) \times H^{-1}(\Omega)$$

is surjective.   ∎

REMARK 13.2   The estimate for T_0 depends only on the geometry. The only condition on T is that

$$T > T_0 = 2R(x^0),$$
$$\text{where } R(x^0) = \sup_{x \in \Gamma}(x - x^0)\nu = \sup_{x \in \Gamma} m_k \nu_k.$$

$2R(x^0)$ can be considered as the diameter of Ω if $x^0 \in \Omega$. By simply replacing x^0 by origo, we get $R(x^0) = \sup_{x \in \Gamma}(x)\nu = x_{max}$, which is the "diameter" of Ω. It is natural to suppose that if our control area Γ_0 becomes larger, T_0 will decrease, and the time T_0 will increase if the control area becomes smaller. This is, for instance, the case if Ω is a convex set, but it is not true in general. ■

REMARK 13.3   Given $T > T_0$, there are an infinite number of controls κ driving the system to rest, since for every $\tilde{T}$ such that

$$T_0 < \tilde{T} < T$$

there is a control $\tilde{\kappa}$ such that $y(\tilde{T}; \tilde{\kappa}) = y'(\tilde{T}; \tilde{\kappa}) = 0$, e.g., one can take $\kappa = \tilde{\kappa}$ in $]0, \tilde{T}[$, $\kappa = 0$ in $]\tilde{T}, T[$. The control κ given by (13.8) is the unique control which minimizes

$$\int_{\Gamma(x^0)} |\kappa|^2 \, d\Gamma dt \tag{13.11}$$

among all the controls driving the system to rest in time T. For a proof see Lions [22] T. 1. pp. 433-440. ■

Before we prove Theorem 13.1 we will state some preliminary lemmas.

13.1.2   Preliminary Lemmas

In this section we give some lemmas which will useful in the proof of Theorem 13.1.

LEMMA 13.2
Let φ be the solution of the system

$$\begin{cases} \varphi'' - \Delta\varphi = 0, \\ \varphi(x,0) = \varphi^0, \; \varphi'(x,0) = \varphi^1 \text{ in } \Omega, \\ \varphi = 0 \text{ on } \Gamma \times]0, T[. \end{cases} \tag{13.12}$$

Then

$$\int_{\Gamma_0 \times]0,T[} \left|\frac{\partial\varphi}{\partial\nu}\right|^2 \, d\Gamma dt \geq c_1(T - T_0)\left[\|\varphi^0\|^2_{H^1_0(\Omega)} + \|\varphi^1\|^2_{L^2(\Omega)}\right] \tag{13.13}$$

and

$$\int_{\Gamma_0 \times]0,T[} \left|\frac{\partial\varphi}{\partial\nu}\right|^2 \, d\Gamma dt \leq c_2(T)\left[\|\varphi^0\|^2_{H^1_0(\Omega)} + \|\varphi^1\|^2_{L^2(\Omega)}\right], \tag{13.14}$$

where $T_0 = 2R(x^0) + \frac{n-1}{\lambda_0}$.

PROOF   We will first show (13.13). We multiply (13.12) by $m_k \frac{\partial \varphi}{\partial x_k}$ and integrate by parts to obtain

$$0 = \int_{\Omega \times]0,T[} \varphi'' m_k \frac{\partial \varphi}{\partial x_k} dx dt - \int_{\Omega \times]0,T[} \Delta \varphi m_k \frac{\partial \varphi}{\partial x_k} dx dt,$$

$$= \left[\int_\Omega \varphi' m_k \frac{\partial \varphi}{\partial x_k} dx \right]_0^T - \int_{\Omega \times]0,T[} \varphi' m_k \frac{\partial \varphi'}{\partial x_k} dx dt$$

$$- \int_{\Gamma \times]0,T[} \frac{\partial \varphi}{\partial x_j} \cos(\angle x_j, \nu) m_k \frac{\partial \varphi}{\partial x_k} d\Gamma dt$$

$$+ \int_{\Omega \times]0,T[} \frac{\partial \varphi}{\partial x_j} \frac{\partial m_k \frac{\partial \varphi}{\partial x_k}}{\partial x_j} dx dt, \tag{13.15}$$

where $(\angle x_k, \nu)$ is the angle between the x_k-direction and the outer normal ν. For simplicity, define

$$X = \left[\int_\Omega \varphi' m_k \frac{\partial \varphi}{\partial x_k} dx \right]_0^T. \tag{13.16}$$

An application of the normal scalar product gives immediately that $\cos(\angle x_k, \nu) = \nu_k$, where ν_k is k'th coordinate of ν. Now, if we use that $\frac{1}{2} \frac{\partial (\varphi')^2}{\partial x_k} = \varphi' \frac{\partial \varphi'}{\partial x_k}$, we get

$$0 = X - \int_{\Omega \times]0,T[} \frac{m_k}{2} \frac{\partial (\varphi')^2}{\partial x_k} dx dt$$

$$- \int_{\Gamma \times]0,T[} \frac{\partial \varphi}{\partial x_j} \nu_j m_k \frac{\partial \varphi}{\partial x_k} d\Gamma dt$$

$$+ \int_{\Omega \times]0,T[} \frac{\partial \varphi}{\partial x_j} \frac{\partial m_k \frac{\partial \varphi}{\partial x_k}}{\partial x_j} dx dt. \tag{13.17}$$

Hence,

$$0 = X - \int_{\Omega \times]0,T[} \frac{m_k}{2} \frac{\partial (\varphi')^2}{\partial x_k} dx dt - \int_{\Gamma \times]0,T[} \frac{\partial \varphi}{\partial \nu} m_k \frac{\partial \varphi}{\partial x_k} d\Gamma dt$$

$$+ \int_{\Omega \times]0,T[} \frac{m_k}{2} \frac{\partial \sum_{j=1}^n (\frac{\partial \varphi}{\partial x_j})^2}{\partial x_k} + \int_{\Omega \times]0,T[} \sum_{j=1}^n \frac{\partial^2 \varphi}{\partial x_j^2} dx dt.$$

This is equivalent to:

$$0 = X - \int_{\Omega \times]0,T[} \frac{m_k}{2} \frac{\partial (\varphi')^2}{\partial x_k} dx dt + \int_{\Omega \times]0,T[} \frac{m_k}{2} \frac{\partial |\nabla \varphi|^2}{\partial x_k} dx dt$$

$$\underbrace{\qquad\qquad\qquad\qquad\qquad\qquad\qquad\qquad\qquad\qquad}_{I}$$

$$- \int_{\Gamma \times]0,T[} \frac{\partial \varphi}{\partial \nu} m_k \frac{\partial \varphi}{\partial x_k} d\Gamma dt \int_{\Omega \times]0,T[} |\nabla \varphi|^2 dx dt. \qquad (13.18)$$

If we consider the integral I we get

$$I = - \int_{\Omega \times]0,T[} \frac{m_k}{2} \frac{\partial (\varphi')^2}{\partial x_k} dx dt + \int_{\Omega \times]0,T[} \frac{m_k}{2} \frac{\partial |\nabla \varphi|^2}{\partial x_k} dx dt$$

$$= \int_{\Omega \times]0,T[} \frac{m_k}{2} \frac{\partial [|\nabla \varphi|^2 - (\varphi')^2]}{\partial x_k} dx dt$$

$$= \int_{\Gamma \times]0,T[} \frac{m_k}{2} [|\nabla \varphi|^2 - (\varphi')^2] \cos(\angle x_k, \nu) d\Gamma dt + \frac{n}{2} \int_{\Omega \times]0,T[} (\varphi')^2 dx dt$$

$$- \frac{n}{2} \int_{\Omega \times]0,T[} |\nabla \varphi|^2 dx dt.$$

Now $\varphi(x, t) = 0$ on $\Gamma \times]0, T[$ and hence so is $\varphi'(x, t) = 0$. Thus,

$$I = \int_{\Gamma \times]0,T[} \frac{m_k}{2} |\nabla \varphi|^2 \cos(\angle x_k, \nu) d\Gamma dt + \frac{n}{2} \int_{\Omega \times]0,T[} (\varphi')^2 dx dt \qquad (13.19)$$

$$- \frac{n}{2} \int_{\Omega \times]0,T[} |\nabla \varphi|^2 dx dt. \qquad (13.20)$$

Hence, (13.18) can be written as

$$0 = X + \frac{n}{2} \int_{\Omega \times]0,T[} (\varphi')^2 dx dt - \int_{\Gamma \times]0,T[} \frac{\partial \varphi}{\partial \nu} m_k \frac{\partial \varphi}{\partial x_k} d\Gamma dt$$

$$+ \int_{\Gamma \times]0,T[} \frac{m_k}{2} \nu_k |\nabla \varphi|^2 d\Gamma dt - \frac{n}{2} \int_{\Omega \times]0,T[} |\nabla \varphi|^2 dx dt$$

$$+ \int_{\Omega \times]0,T[} |\nabla \varphi|^2 dx dt,$$

which is equivalent to

$$0 = X + \frac{n}{2} \int_{\Omega \times]0,T[} (\varphi')^2 dxdt - \frac{1}{2} \int_{\Omega \times]0,T[} (\varphi')^2 dxdt$$

$$- \int_{\Gamma \times]0,T[} \frac{\partial \varphi}{\partial \nu} m_k \frac{\partial \varphi}{\partial x_k} d\Gamma dt + \int_{\Gamma \times]0,T[} \frac{m_k}{2} \nu_k |\nabla \varphi|^2 d\Gamma dt$$

$$- \frac{n}{2} \int_{\Omega \times]0,T[} |\nabla \varphi|^2 dxdt + \frac{1}{2} \int_{\Omega \times]0,T[} |\nabla \varphi|^2 dxdt$$

$$+ \frac{1}{2} \int_{\Omega \times]0,T[} |\nabla \varphi|^2 dxdt$$

$$+ \frac{1}{2} \int_{\Omega \times]0,T[} (\varphi')^2 dxdt$$

$$= X + \frac{n-1}{2} \int_{\Omega \times]0,T[} [(\varphi')^2 - |\nabla \varphi|^2] dxdt - \int_{\Gamma \times]0,T[} \frac{\partial \varphi}{\partial \nu} m_k \frac{\partial \varphi}{\partial x_k} d\Gamma dt$$

$$+ \int_{\Gamma \times]0,T[} \frac{m_k}{2} \nu_k |\nabla \varphi|^2 d\Gamma dt + \frac{1}{2} \int_{\Omega \times]0,T[} [(\varphi')^2 + |\nabla \varphi|^2] dxdt.$$

Note that

$$\frac{\partial \varphi}{\partial x_k} = \nu_k \left(\frac{\partial \varphi}{\partial \nu} \right),$$

$$\frac{\partial \varphi}{\partial \nu} m_k \frac{\partial \varphi}{\partial x_k} = m_k \nu_k \left(\frac{\partial \varphi}{\partial \nu} \right)^2,$$

$$\frac{m_k \nu_k}{2} |\nabla \varphi|^2 = \frac{m_k \nu_k}{2} \left(\frac{\partial \varphi}{\partial \nu} \right)^2.$$

Hence,

$$0 = X + \frac{n-1}{2} \int_{\Omega \times]0,T[} [(\varphi')^2 - |\nabla \varphi|^2] dxdt$$

$$- \int_{\Gamma \times]0,T[} \frac{m_k}{2} \nu_k \left(\frac{\partial \varphi}{\partial \nu} \right)^2 d\Gamma dt + \frac{1}{2} \int_{\Omega \times]0,T[} [(\varphi')^2 + |\nabla \varphi|^2] dxdt.$$

Let

$$Y = \int_{\Omega \times]0,T[} [(\varphi')^2 - |\nabla \varphi|^2] dxdt$$

and define the energy, which is constant (see Treves [29] pp. 116), as

$$E(t) = E_0 = \frac{1}{2} \int_{\Omega} [(\varphi')^2 + |\nabla \varphi|^2] dx. \qquad (13.21)$$

Hence,

$$TE_0 - \int_{\Gamma \backslash \Gamma_0 \times]0,T[} \frac{m_k}{2} \nu_k \left(\frac{\partial \varphi}{\partial \nu}\right)^2 d\Gamma dt = -X - \frac{n-1}{2} Y$$

$$+ \int_{\Gamma_0 \times]0,T[} \frac{m_k}{2} \nu_k \left(\frac{\partial \varphi}{\partial \nu}\right)^2 d\Gamma dt.$$

$$(13.22)$$

Now, since $-m_k \nu_k \geq 0$ on $\Gamma \backslash \Gamma_0$, the integral on the left-hand side is positive. If we define

$$R(x^0) = \sup_{x \in \Gamma} m_k \nu_k \qquad (13.23)$$

we get that

$$TE_0 \leq |X| + \frac{n-1}{2} |Y| + \frac{R(x^0)}{2} \int_{\Gamma_0 \times]0,T[} \left|\frac{\partial \varphi}{\partial \nu}\right|^2 d\Gamma dt. \qquad (13.24)$$

In order to obtain the inequality (13.13), we need the following two inequalities. Using the fact that

$$\|\varphi'\|_{L^2(\Omega)} \|\nabla \varphi\|_{L^2(\Omega)} \leq \frac{1}{2} \left[\|\varphi'\|^2_{L^2(\Omega)} + \|\nabla \varphi\|^2_{L^2(\Omega)}\right] = E_0,$$

and that there exists a constant λ_0 (see Poincaré's inequality) such that

$$\lambda_0 \|\varphi\|_{L^2(\Omega)} \leq \|\nabla \varphi\|_{L^2(\Omega)} \text{ for all } \varphi \in H_0^1(\Omega),$$

we obtain

$$2 \left|\int_\Omega \varphi' m_k \frac{\partial \varphi}{\partial x_k} dx\right| \leq 2R(x^0) \|\varphi'\|_{L^2(\Omega)} \|\nabla \varphi\|_{L^2(\Omega)} \leq 2R(x_k) E_0.$$

$$\left|\int_\Omega \varphi' \varphi \, dx\right| \leq \|\varphi'\|_{L^2(\Omega)} \frac{1}{\lambda_0} \|\nabla \varphi\|_{L^2(\Omega)} \leq \frac{1}{\lambda_0} E_0.$$

Because of this and Lemma (13.3), we can now write (13.24) as

$$TE_0 \leq \left(2R(x^0) + \frac{n-1'}{\lambda_0}\right) E_0 + \frac{R(x^0)}{2} \int_{\Gamma_0 \times]0,T[} \left|\frac{\partial \varphi}{\partial \nu}\right|^2 d\Gamma dt, \qquad (13.25)$$

which is equivalent to

$$\int_{\Gamma_0 \times]0,T[} \left(\frac{\partial \varphi}{\partial \nu}\right)^2 d\Gamma dt \geq \frac{2}{R(x^0)} (T - T_0) E_0, \qquad (13.26)$$

$$\text{where } T_0 = 2R(x^0) + \frac{n-1}{\lambda_0}. \qquad (13.27)$$

This shows the inequality (13.13). The second inequality (13.14) is proved by taking $F = g = 0$ in (12.157) in Theorem 12.15, Section 12.1.2, Chapter 12. ∎

LEMMA 13.3
Let φ be the solution of the system (13.12). Then

$$Y = \int_{\Omega \times]0,T[} [|\varphi'|^2 - (\nabla\varphi)^2] dx dt = \left[\int_\Omega \varphi\varphi' dx\right]_0^T. \qquad (13.28)$$

PROOF
Integration by parts gives

$$\int_{\Omega \times]0,T[} \varphi'\varphi' dx dt = \left[\int_\Omega \varphi\varphi' dx\right]_0^T - \int_{\Omega \times]0,T[} \varphi''\varphi dx dt.$$

Hence,

$$\left[\int_\Omega \varphi\varphi' dx\right]_0^T = \int_{\Omega \times]0,T[} |\varphi'|^2 dx dt + \underbrace{\int_{\Omega \times]0,T[} \varphi\Delta\varphi dx dt}_{I_1} .$$

If we take $v = \nabla\varphi$ in

$$\int_\Omega div(\varphi v) dx = \int_\Gamma \varphi\langle v, \nu\rangle d\Gamma,$$

we get that

$$\int_{\Omega \times]0,T[} [\varphi\Delta\varphi dt dx + |\nabla\varphi|^2] dx dt = \int_{\Gamma \times]0,T[} \varphi\frac{\partial\varphi}{\partial\nu} d\Gamma dt.$$

From the boundary conditions we obtain that the R.H.S. is zero, so

$$I_1 = \int_{\Omega \times]0,T[} \varphi\Delta\varphi dt dx = -\int_{\Omega \times]0,T[} |\nabla\varphi|^2 dx dt.$$

∎

LEMMA 13.4
Let φ be the solution of

$$\begin{cases} \varphi'' - \Delta\varphi = 0, \\ \varphi(0, x) = \varphi^0 , \ \varphi'(0, x) = \varphi^1 \text{ in } \Omega, \\ \varphi = 0 \text{ on } \Gamma \times]0, T[. \end{cases} \qquad (13.29)$$

Then

$$\langle \Lambda\{\varphi^0, \varphi^1\}, \{\varphi^0, \varphi^1\}\rangle = \int_{\Gamma_0 \times]0,T[} \left|\frac{\partial\varphi}{\partial\nu}\right|^2 d\Gamma dt. \qquad (13.30)$$

PROOF
We multiply (13.29) by φ and integrate by parts to get

$$\begin{aligned}
0 &= \int_{\Omega \times]0,T[} [\varphi\psi'' - \varphi\Delta\psi]dtdx \\
&= \int_{\Omega} [\varphi\psi']_0^T dx - \int_{\Omega \times]0,T[} [\varphi'\psi' - \varphi\Delta\psi]dtdx \\
&= \int_{\Omega} \left(-\varphi^0\psi^1 - [\psi\varphi']_0^T\right) dx + \int_{\Omega \times]0,T[} [\varphi''\psi - \varphi\Delta\psi]dtdx \\
&= -\langle \Lambda\{\varphi^0, \varphi^1\}, \{\varphi^0, \varphi^1\}\rangle + \int_{\Omega \times]0,T[} [\psi\Delta\varphi - \varphi\Delta\psi]dtdx \\
&= -\langle \Lambda\{\varphi^0, \varphi^1\}, \{\varphi^0, \varphi^1\}\rangle + \int_{\Gamma \times]0,T[} \left[\psi\frac{\partial\psi}{\partial\nu} - \varphi\frac{\partial\psi}{\partial\nu}\right] d\Gamma dt,
\end{aligned}$$

where the last equality is obtained by an application of the modified Green's formula as seen in (11.27). Using the boundary condition, we finally get the result (13.30)   ∎

PROOF OF THEOREM 13.1:
From Lemma 13.4 we obtain

$$\langle\{\psi'(0), -\psi(0)\}, \{\varphi^0, \varphi^1\}\rangle = \langle\Lambda\{\varphi^0, \varphi^1\}, \{\varphi^0, \varphi^1\}\rangle \qquad (13.31)$$

$$= \int_{\Omega} \{\psi'(0)\varphi^0 - \psi(0)\varphi^1\}dx \qquad (13.32)$$

$$= \int_{\Gamma_0 \times]0,T[} \left|\frac{\partial\varphi}{\partial\nu}\right|^2 d\Gamma dt. \qquad (13.33)$$

The main point is to show that

$$\langle\Lambda\{\varphi^0, \varphi^1\}, \{\varphi^0, \varphi^1\}\rangle = \int_{\Gamma_0 \times]0,T[} \left|\frac{\partial\varphi}{\partial\nu}\right|^2 d\Gamma dt \qquad (13.34)$$

defines a norm on the set of initial data $\{\varphi^0, \varphi^1\}$. Then we show that this norm (13.34) is equivalent to the usual norm of $H_0^1(\Omega) \times L^2(\Omega)$ and infer that Λ is an isomorphism from $H_0^1(\Omega) \times L^2(\Omega)$ onto $H^{-1}(\Omega) \times L^2(\Omega)$.

We divide the proof into three steps.

STEP 1.

In order to prove (13.34) defines a norm and hereafter the equivalence of the norms, we show that there exist two constants c_1 and $c_2(T)$ such that the following two inequalities hold:

$$\int_{\Gamma_0 \times]0,T[} \left|\frac{\partial\varphi}{\partial\nu}\right|^2 d\Gamma dt \geq c_1 (T - T_0) \left[\|\varphi^0\|^2_{H^1_0(\Omega)} + \|\varphi^1\|^2_{L^2(\Omega)}\right] \qquad (13.35)$$

and

$$\int_{\Gamma_0 \times]0,T[} \left|\frac{\partial\varphi}{\partial\nu}\right|^2 d\Gamma dt \leq c_2(T) \left[\|\varphi^0\|^2_{H^1_0(\Omega)} + \|\varphi^1\|^2_{L^2(\Omega)}\right]. \qquad (13.36)$$

(13.35) and (13.36) follow directly from Lemma 13.2.

STEP 2. *(13.34) is a norm.*

We only have to prove that:

$$\langle\Lambda\{\varphi^0, \varphi^1\}, \{\varphi^0, \varphi^1\}\rangle = \int_{\Gamma_0 \times]0,T[} \left|\frac{\partial\varphi}{\partial\nu}\right|^2 d\Gamma dt = 0 \Leftrightarrow \varphi = 0. \qquad (13.37)$$

$\langle\Lambda\{\varphi^0, \varphi^1\}, \{\varphi^0, \varphi^1\}\rangle$ is clearly a self-adjoint and nonnegative operator for $T > T_0$. So for $T > T_0$ (13.37) follows immediately, because then

$$\langle\Lambda\{\varphi^0, \varphi^1\}, [\varphi^0, \varphi^1]\rangle \geq a\|\{\varphi^0, \varphi'\}\|^2_{H^1_0(\Omega) \times L^2(\Omega)}, a \in R_+. \qquad (13.38)$$

Hence (13.34) defines a norm. It is clear from (13.35) and (13.36) that for $T > T_0$, the norm (13.34) is equivalent to the usual norm of $H^1_0(\Omega) \times L^2(\Omega)$.

STEP 3. Λ *is an isomorphism.*

We need to show that Λ is an isomorphism from $H^1_0(\Omega) \times L^2(\Omega)$ onto $H^{-1}(\Omega) \times L^2(\Omega)$. We have that $\langle\Lambda\{\tilde{\varphi}^0, \tilde{\varphi}^1\}, \{\varphi^0, \varphi^1\}\rangle$ defines a scalar product on the set of initial data. Notice now that the norm (13.34) is defined by this scalar product and defines a Hilbert space on the set of initial data, which is equivalent to the Hilbert space $H^1_0(\Omega) \times L^2(\Omega)$. By Riesz Representation Theorem we know that there exists an isomorphic mapping K between $F = H^1_0(\Omega) \times L^2(\Omega)$ and its dual $F' = H^{-1}(\Omega) \times L^2(\Omega)$ defined by

$$\langle u, v\rangle_F = \langle Ku, v\rangle_{F',F}. \qquad (13.39)$$

Writing this in an another way

$$\langle\Lambda u, v\rangle = \langle Ku, v\rangle_{F',F}. \qquad (13.40)$$

We conclude that Λ is isomorphic mapping of $H^1_0(\Omega) \times L^2(\Omega)$ onto $H^{-1}(\Omega) \times L^2(\Omega)$.

The proof of Theorem 13.1 is now completed.

REMARK 13.4   From the proof of Theorem 13.1, it is seen that the estimate of T_0, namely $T_0 = 2R(x^0) + \frac{n-1}{\lambda_0}$, does not agree with the estimate stated in Remark 13.2. The latter is preferable because it gives an estimate of the time T_0 that only depends on the geometry of Ω(i.e., the "diameter" of Ω). The estimate $T_0 = 2R(x^0)$ is given in Proposition 13.5 below.   ∎

PROPOSITION 13.5
If we follow the notation of Theorem 13.1, there exists a constant c such that

$$\int_{\Gamma_0 \times]0,T[} \left| \frac{\partial \varphi}{\partial \nu} \right|^2 d\Gamma dt \geq c(T - 2R(x^0))E_0. \tag{13.41}$$

PROOF   Using a compactness argument and the uniqueness property that

$$\varphi'' - \Delta\varphi = 0 \text{ in } \Omega \times]0,T[,$$
$$\varphi = 0 \text{ on } \Gamma \times]0,T[,$$
$$\frac{\partial \varphi}{\partial \nu} = 0 \text{ on } \Gamma(x^0) \times]0,T[,$$
$$T > 2R(x^0)$$

implies $\varphi = 0$. The proof can be found in Komornik [13].
We can also obtain the inequality (13.41) by using an alternative estimate for

$$\left| X + \frac{(n-1)}{2}Y \right|$$

in the proof of Theorem 13.1 as compared to the one previously used. Let us set

$$\xi(t) = \int_\Omega \varphi' \left(m_k \frac{\partial \varphi}{\partial x_k} + \frac{(n-1)}{2}\varphi \right) dx. \tag{13.42}$$

Using the definitions for X and Y we have

$$\left| X + \frac{(n-1)}{2}Y \right| = \left| \left[\int_\Omega \varphi' m_k \frac{\partial \varphi}{\partial x_k} dx \right]_0^T + \left[\frac{(n-1)}{2} \int_\Omega \varphi\varphi' dx \right]_0^T \right| = \left| [\xi]_0^T \right|.$$

Hence, it is readily seen that

$$\left| X + \frac{(n-1)}{2}Y \right| \leq | \xi(T) | + | \xi(0) |. \tag{13.43}$$

Equation (13.42) can be estimated by Young's Inequality.

$$\xi(t) = \int_\Omega \varphi' \left(m_k \frac{\partial \varphi}{\partial x_k} + \frac{(n-1)}{2} \varphi \right) dx$$

$$= \int_\Omega (R(x^0)\varphi') \frac{1}{R(x^0)} \left(m_k \frac{\partial \varphi}{\partial x_k} + \frac{(n-1)}{2} \varphi \right) dx$$

$$\leq \frac{1}{2} R(x^0) \|\varphi'\|_{L^2(\Omega)}^2 + \frac{1}{2R(x^0)} \left\| m_k \frac{\partial \varphi}{\partial x_k} + \frac{(n-1)}{2} \varphi \right\|_{L^2(\Omega)}^2 \qquad (13.44)$$

We observe that

$$\left\| m_k \frac{\partial \varphi}{\partial x_k} + \frac{(n-1)}{2} \varphi \right\|_{L^2(\Omega)}^2 = \int_\Omega \left| m_k \frac{\partial \varphi}{\partial x_k} + \frac{(n-1)}{2} \varphi \right|^2 dx$$

$$= \left\| m_k \frac{\partial \varphi}{\partial x_k} \right\|_{L^2(\Omega)}^2 + \left(\frac{n-1}{2} \right)^2 \|\varphi\|_{L^2(\Omega)}^2 +$$

$$\underbrace{\frac{(n-1)}{2} \int_\Omega m_k \varphi \frac{\partial \varphi}{\partial x_k} dx + \frac{(n-1)}{2} \int_\Omega m_k \varphi \frac{\partial \varphi}{\partial x_k} dx}_{I_1}.$$

We can evaluate I_1 in the following manner

$$I_1 = \frac{(n-1)}{2} \int_\Gamma m_k \nu_k \mid \varphi \mid^2 d\Gamma - \frac{(n-1)}{2} \int_\Omega \varphi \left(n\varphi + m_k \frac{\partial \varphi}{\partial x_k} \right) dx$$

$$+ \frac{(n-1)}{2} \int_\Gamma m_k \nu_k \mid \varphi \mid^2 d\Gamma - \frac{(n-1)}{2} \int_\Omega \varphi \left(n\varphi + m_k \frac{\partial \varphi}{\partial x_k} \right) dx.$$

Since $\varphi = 0$ on $\Gamma \times]0, T[$, we now have

$$I_1 = \frac{-n(n-1)}{2} \|\varphi\|_{L^2(\Omega)}^2. \qquad (13.45)$$

Since

$$\left(\frac{n-1}{2} \right)^2 - \frac{n(n-1)}{2} \leq 0$$

and

$$\left\| m_k \frac{\partial \varphi}{\partial x_k} \right\|_{L^2(\Omega)}^2 \leq R(x^0)^2 \|\nabla \varphi\|_{L^2(\Omega)}^2,$$

we can conclude that

$$\left\| m_k \frac{\partial \varphi}{\partial x_k} + \frac{(n-1)}{2} \varphi \right\|_{L^2(\Omega)}^2 \leq R(x^0)^2 \|\nabla \varphi\|_{L^2(\Omega)}^2. \qquad (13.46)$$

Applying (13.46) in inequality (13.44) we have

$$|\,\xi(t)\,| \leq R(x^0)E_0. \qquad (13.47)$$

From the inequality (13.43) we have

$$\left| X + \frac{(n-1)}{2}Y \right| \leq 2R(x^0)E_0. \qquad (13.48)$$

Applying this in equation (13.24), we get the inequality (13.41). ∎

13.1.3 A Discussion of the Control Area

In the discussion of the choice of control area, there are basically two main points of interest:

- assumptions on the geometry of Ω in the choice of a control area

- the dependency of the time required to bring the system to rest $T(x^0)$ on the choice of the control area

These two assertions will be discussed in the following. In Theorem 13.1 we demand that $T > 2R(x^0)$ in the inequality (13.41) in order to guarantee that the norm (13.34) is equivalent to the usual norm in $H_0^1(\Omega) \times L^2(\Omega)$. From this we conclude that the minimal time required to achieve exact controllability $T(x^0)$ is $2R(x^0)$. $R(x^0)$ is determined by the geometry of Ω and the choice of x^0 alone. By varying x^0, the control area Γ_0 changes too, and these changes depend on the geometrical nature of Ω. In choosing Γ_0 we have the following two options. We can either minimize the time taken to bring the system to rest, i.e., choose

$$x^0 = \{x^0 \in R^n \mid R(x^0) = \min_{x^0 \in R^n} R(x^0)\} \qquad (13.49)$$

and perhaps pay the price by having to place controls on a larger part of Γ or use a stronger force, or opt for the converse situation, i.e., choose the control area better suited to the needs (this can either be from a practical point of view where we have some kind of an "ideal" for $\Gamma(x^0)$ and we choose an x^0 that best approximates this $\Gamma(x^0)$) at hand and accept that it will perhaps take a longer time to bring the system to rest.

Notice that we use the word "perhaps" has been used. Acting on a larger part of the boundary does not automatically imply T_0 will be reduced. This depends upon the nature of Ω, i.e., the concavity or convexity of the part of the boundary considered.

Let us assume for the moment that Ω is a convex set. It is easily seen that $\Gamma_0 = \Gamma$. The best choice is obviously $R(x^0) =$ diameter of Ω. If x^0 is chosen to be outside Ω, we see that Γ_0 becomes a smaller set if $R(x^0)$

becomes larger. Intuitively, this quite logical, since we have a convex set. It takes a longer time to drive the system to rest.

In the selection of the control area, it will be natural to ask if the area can be chosen arbitrarily. More precise: given a specific part of the boundary, say $\Gamma_1 \subset \Gamma$, Γ_1 not of the type $\Gamma(x^0)$, can we *always* find $x^0 \in R^n$ such that $\Gamma_1 = \Gamma(x^0)$? The answer is no. Once we have chosen $x^0 \in R^n$ arbitrarily, the control area $\Gamma(x^0)$ is fixed and the control given by HUM will act on the entire boundary $\Gamma(x^0)$. The converse is not true. We cannot generally choose our control area in advance arbitrarily and then find a suitable $x^0 \in R^n$.

We can, however, formulate a question related to the above in the following manner: Given a specific part of the boundary, say $\Gamma_1 \subset \Gamma$, Γ_1 not of the type $\Gamma(x^0)$, we are able to find a control κ of the following type

$$y = \begin{cases} \kappa \text{ on } \Gamma_1 \times]0, T[\\ 0 \text{ on } (\Gamma \setminus \Gamma_1) \times]0, T[\end{cases} \qquad (13.50)$$

Note that by simply choosing $x^0 \in R^n$ such that $\Gamma_1 \subset \Gamma_0$ and then by setting $\kappa = 0$ on $\Gamma_0 \setminus \Gamma_1$, we have a special case of this second question. HUM does (at least in principle) give a method to find the required control in this case. There is, however, a problem. If we consider the system of Theorem 13.1 the proof of the theorem will have to be modified from (13.22) onward. This is due to the fact that the estimates used here would no longer be valid and alternative ones would have to be found. Whether this is really possible using the method of proof presented here remains an open question.

REMARK 13.5   According to Triggiani [30], the same exact controllability result as in Theorem 13.1 can be obtained by acting on a strictly smaller subset Γ_1 of the boundary than Γ_0. The disadvantage, however, is that the definition of the subset Γ_1 is not as straightforward as the definition of Γ_0 (Definition 13.1).   ∎

13.1.4   Exact Controllability in Polygons and Polyhedra

We have until now studied the problem of exact controllability in smooth domains. The problem, however, becomes tricky when the domain Ω has corners. Grisvard [8] provides us, by an application of HUM, some exact controllability results for systems whose evolution is determined by the classical wave equation in polygons and polyhedra. The aim of this section is only to state some of these results. The reader can consult Grisvard [8] and [9] for a more comprehensive presentation of the problem of exact controllability in nonsmooth domains. We also refer the reader to Grisvard

[7] for a general introduction of Sobolev spaces in nonsmooth domains and to elliptic boundary value problems in this setting.

Geometrical Assumptions:

Let Ω_p be an open, bounded, connected set in R^n where $n = 2$ or $n = 3$. Let

$$\Gamma = \bigcup_{j=1}^{J} \Gamma_j. \tag{13.51}$$

be the boundary of Ω_p, where $\Gamma_j, 1 \le j \le J$ denotes segments in the case $n = 2$ and planes when $n = 3$. Ω_p is thus a polygon for $n = 2$ and a polyhedron for $n = 3$. We also assume that Ω_p lies locally on one side of Γ. Let ν_j denote the unit outward normal to Γ_j.

In order to provide a picture of the types of regularity results that can be obtained in nonsmooth domains, we present some regularity results for elliptic as well as for hyperbolic Dirichlet problems. As in the case of smooth domains, these regularity results are important in obtaining exact controllability results.

THEOREM 13.6

There exists an $s \in]3/2, 2]$ such that every $\Phi \in H_0^1(\Omega_p)$ which is a solution of $\Delta\Phi = f \in L^2(\Omega_p)$ satisfies $\Phi \in H^s(\Omega_p)$. Moreover, $s = 2$ if Ω_p is a convex set.

A (sketch of the) proof can be found in Grisvard [8], p. 368.   ∎

The theorem above plays a major role in obtaining some identities in which the normal derivative is involved. These are in turn used to obtain *a priori* estimates.

We also have the following regularity result in polygons and polyhedra where the evolution of the system is described by the classical wave equation. Note that the interior regularity result is similar to the one obtained in smooth domains in Theorem 3, Chapter 12.

THEOREM 13.7

Consider the following system:

$$\begin{cases} \Phi'' - \Delta\Phi = 0 \text{ in } \Omega_p \times]0, T[, \\ \Phi(0) = \Phi^0, \ , \Phi'(0) = \Phi^1, \\ \Phi = g \text{ on } \Gamma \times]0, T[\end{cases} \tag{13.52}$$

where

$$\begin{cases} \Phi^0 \in L^2(\Omega_p), \Phi^1 \in H^{-1}(\Omega_p), \\ g \in L^2(\Gamma \times]0, T[). \end{cases} \tag{13.53}$$

Then the weak solution of (13.52) satisfies

$$\Phi \in L^\infty(0, T; L^2(\Omega)). \tag{13.54}$$

This weak solution is given by

$$\int_{\Omega_p \times]0, T[} \Phi f dx dt = \int_\Omega \Phi^1 \varphi^0 dx - \int_\Omega \Phi^0 \varphi^1 dx - \sum_{j=1}^J \int_{\Gamma_j \times]0, T[} g \frac{\partial \varphi}{\partial \nu_j} d\Gamma dt, \tag{13.55}$$

where $\varphi \in C([0, T]; H_0^1(\Omega_p)) \cap C^1([0, T]; L^2(\Omega_p))$ satisfies

$$\begin{cases} \varphi'' - \Delta \varphi = f \text{ in } \Omega_p \times]0, T[, \\ \varphi(T) = \varphi'(T) = 0 \text{ in } \Omega_p \times]0, T[, \\ \varphi = 0 \text{ on } \Gamma \times]0, T[\end{cases} \tag{13.56}$$

for any $f \in L^1(0, T; L^2(\Omega))$.

Application of HUM:
Let $x^0 \in R^n$ an arbitrary point. Let

$$m(x) = x - x^0. \tag{13.57}$$

Let J^* denote the set of indices j for which $m \cdot \nu_j \geq 0$ for every $x \in \Gamma_j$, where ν_j is the unit outward normal to Γ_j. Let

$$\Gamma_0 = \bigcup_{j \in J^*} \Gamma_j. \tag{13.58}$$

We can now state the following exact controllability result:

THEOREM 13.8
Assume that Ω_p satisfies the geometrical hypotheses stated in Section 13.1.4. Then for any couple $\{y^0, y^1\} \in L^2(\Omega_p) \times H^{-1}(\Omega_p)$ there exists a control $\kappa \in L^2(\Gamma_0 \times]0, T[)$ and T_0 such that for every $T > T_0$ the solution y of the system given by

$$\begin{cases} y'' - \Delta y = 0 \text{ in } \Omega_p \times]0, T[, \\ y = \kappa \text{ on } \Gamma_0 \times]0, T[, \\ y = 0 \text{ on } \Gamma \backslash \Gamma_0 \times]0, T[, \\ y(0) = y^0, \ y'(0) = y^1 \text{ in } \Omega_p \end{cases} \tag{13.59}$$

satisfies

$$y(T) = y'(T) = 0. \tag{13.60}$$

We refer the reader to Grisvard [8] for the full proof of the theorem. A few comments are in order, however.

The method of the proof is based directly on HUM. As usual, we have the two auxiliary coupled φ and ψ systems:

$$\begin{cases} \varphi'' - \Delta\varphi = 0, \ \text{in} \ \Omega_p \times]0, T[, \\ \varphi = 0 \ \text{on} \ \Gamma \times]0, T[, \\ \varphi(0) = \varphi^0, \ \ \varphi'(0) = \varphi^1 \end{cases} \tag{13.61}$$

and

$$\begin{cases} \psi'' - \Delta\psi = 0, \ \text{in} \ \Omega_p \times]0, T[, \\ \psi = \frac{\partial\varphi}{\partial\nu_j} \ \text{on} \ \Gamma_0 \times]0, T[, \\ \psi = 0 \ \text{on} \ \Gamma\backslash\Gamma_0 \times]0, T[, \\ \psi(T) = \varphi'(T) = 0. \end{cases} \tag{13.62}$$

It can be shown that with appropriate assumptions on the data, there exist unique solutions for the systems above. By various *a priori* estimates it is also possible to show that

$$c_1(\|\varphi^0\|_{H_0^1(\Omega_p)} + \|\varphi^1\|_{L^2(\Omega_p)}) \leq \sum_{j \in J_*} \int_{\Gamma \times]0, T[} \left(\frac{\partial\varphi}{\partial\nu}\right)^2 d\Gamma dt \tag{13.63}$$

$$\leq c_2(\|\varphi^0\|_{H_0^1(\Omega_p)} + \|\varphi^1\|_{L^2(\Omega_p)}), \tag{13.64}$$

where c_1 and c_2 are positive constants. Now, using the duality principle of HUM, it is possible to obtain the result stated in Theorem 13.8.

13.2 The Hilbert Uniqueness Method

In the preceeding section we saw how HUM could be applied in the case of the classical wave equation with Dirichlet boundary control. We will now introduce HUM for more general systems by extending the considerations made in the previous section. HUM deals with the problem of exact controllability, and the systems considered are those governed by propagation equations with boundary control.

We consider the system :

$$y'' + Ay = 0 \ \text{in} \ \Omega \times]0, T[. \tag{13.65}$$

with the initial conditions:

$$y(0) = y^0, \quad y'(0) = y^1 \quad \text{in } \Omega. \tag{13.66}$$

We assume that A is a symmetric elliptic operator of order $2m$.

We act on the boundary of the system in the following manner:

$$B_j y = \kappa_j \text{ on } \Gamma \times \,]0, T[\,; \; 1 \le j \le m, \tag{13.67}$$

where B_j are the boundary operators such that Problem (13.65)-(13.67) is well posed in appropriate function spaces. An important question is: What are the necessary and sufficient conditions on A and B_j for this to be true? This question has been taken up in Chapter 12 where the basic theory of hyperbolic mixed problems is presented.

13.2.1 The Problem of Exact Controllability

The problem of exact controllability in linear systems is the following :

- Given $T > 0$, is it possible to find for any couple $\{y^0, y^1\}$, controls κ_j such that if $y(\kappa)$ denotes the solution to Problem (13.65)-(13.67), then

$$y(T; \kappa) = y'(T; \kappa) = 0? \tag{13.68}$$

REMARK 13.6   The following two points are of some interest:

- In practice it is not always possible to act on the entire boundary, but only on parts of it. In the model above, this corresponds to imposing the following constraints on the boundary controls κ_j,

$$\kappa_j = 0 \quad \text{on } \Gamma \backslash \Gamma_0,$$

i.e., we act only on the part Γ_0 of the boundary Γ.

- We can also have a situation where we only use some of the boundary operators, i.e., $B_j y = \kappa_j$ for $j \in J \subset [1, .., m]$ and $B_j y = 0$ for $\forall j \notin J$.

∎

We will now describe the application of HUM on the system (13.65) - (13.67).

Let us for a moment consider the following problem:

$$\begin{cases} \varphi'' + A\varphi = 0, \\ \varphi(0) = \varphi^0 \,, \varphi'(0) = \varphi^1, \\ B_j \varphi = 0; \; 1 \le j \le m. \end{cases} \tag{13.69}$$

Since our goal is to satisfy equation (13.68), our original problem corresponds to the following backward problem of (13.69).

$$\begin{cases} \psi'' + A\psi = 0, \\ \psi(x,T) = \psi'(x,T) = 0, \\ B_j\psi = \kappa_j, \ 1 \le j \le m. \end{cases} \quad (13.70)$$

From Chapter 12 we have results pertaining to the existence, uniqueness, and regularity of the solutions to the systems above and the traces thereof.

Let us now assume $\kappa_j = -C_j\varphi$, $1 \le j \le m$ in (13.70) where $C'_j s$ are the complementary boundary operators of the $B'_j s$ found by the modified Green's formula. We have, therefore, the following system

$$\begin{cases} \psi'' + A\psi = 0, \\ \psi(x,T) = \psi'(x,T) = 0, \\ B_j\psi = -C_j\varphi, \ 1 \le j \le m. \end{cases} \quad (13.71)$$

Notice that the systems (13.69) and (13.71) are "coupled" through the boundary data.

13.2.2 The Operator Λ

Our aim is now to characterize this coupling between (13.69) and (13.71). This is where the operator Λ plays an important role.

If both the systems (13.69) and (13.71) admit unique solutions, we can then define an operator Λ which acts on the couple $\{\varphi^0, \varphi^1\}$ by

$$\Lambda\{\varphi^0, \varphi^1\} = \{\psi'(0), -\psi(0)\} \quad (13.72)$$

since φ is used as a boundary condition in (13.71). If we can find spaces where Λ is invertible for T large enough, then by solving for any given couple $\{y^1, y^0\}$ the following equation

$$\Lambda\{\varphi^0, \varphi^1\} = \{y^1, -y^0\}, \quad (13.73)$$

we obtain the initial values in (13.69), and this system can be solved. Since we have chosen $\kappa_j = -C_j\varphi$ where φ is the solution of (13.69) and $\{\varphi^0, \varphi^1\}$ fulfills (13.73), we have

$$y(\kappa) = \psi. \quad (13.74)$$

It is now clear that (13.74) fulfills (13.68) by default since (13.71) is fulfilled, i.e., we have succeeded in finding a control κ that drives the system to rest at time T.

We will now identify the sufficient condition on the space in which the couple $\{y^0, y^1\}$ can lie so that the method described above can apply.

In order to do this, we are in need of the identity given in the example below.

Example 13.1

The following identity holds:

$$\langle \Lambda\{\varphi^0, \varphi^1\}, \{\varphi^0, \varphi^1\}\rangle = \sum_{j=1}^{m} \int_{\Gamma \times]0,T[} \mid C_j \varphi \mid^2 d\Gamma dt. \tag{13.75}$$

PROOF   Multiplying (13.69) by φ and integrating by parts over $\Omega \times]0, T[$ we have

$$0 = \int_{\Omega \times]0,T[} (\psi'' + A\psi)\varphi dx dt$$

$$= \left[\int_\Omega \psi' \varphi dx\right]_0^T - \int_{\Omega \times]0,T[} \psi' \varphi' dx dt + \int_{\Omega \times]0,T[} (A\psi)\varphi dx dt$$

$$= -\int_\Omega \psi'(0)\varphi(0)dx - \left[\int_\Omega \psi\varphi' dx\right]_0^T$$

$$+ \int_{\Omega \times]0,T[} \psi\varphi'' dx dt + \int_{\Omega \times]0,T[} (A\psi)\varphi dx dt$$

$$= -\int_\Omega \psi'(0)\varphi(0)dx + \int_\Omega \psi(0)\varphi'(0)dx$$

$$+ \int_{\Omega \times]0,T[} \psi\varphi'' dx dt + \int_{\Omega \times]0,T[} (A\psi)\varphi dx dt.$$

Using the facts that

$$\langle\{\psi'(0), -\psi(0)\}\}\rangle = \langle \Lambda\{\varphi^0, \varphi^1\}, \{\varphi^0, \varphi^1\}\rangle$$

$$= \int_\Omega \{\psi'(0)\varphi^c - \psi(0)\varphi^1\} dx$$

and $\varphi'' = A\varphi$ in the result above, we get

$$\langle \Lambda\{\varphi^0, \varphi^1\}, \{\varphi^0, \varphi^1\}\rangle = \int_{\Omega \times]0,T[} (A\psi)\varphi dx dt - \int_{\Omega \times]0,T[} \psi(A\varphi)dx dt.$$

$$\tag{13.76}$$

Equation (11.27) derived in Example 11.2 gives the following equation:

$$\int_\Omega ((A\psi)\varphi - (A\varphi)\psi)dx = \sum_{j=0}^{m-1} \int_\Gamma ((C_j\varphi)(B_j\psi) - (C_j\psi)(B_j\varphi))d\Gamma. \quad (13.77)$$

Using the equation above and the fact that $B_j\varphi = 0$, we now have

$$\langle \Lambda\{\varphi^0, \varphi^1\}, \{\varphi^0, \varphi^1\}\rangle = \sum_{j=1}^{m} \int_{\Gamma\times]0,T[} | C_j\varphi |^2 \, dt. \quad (13.78)$$

The idea of proving identity (13.75) is the following:

Let us assume that the following equation

$$\left(\sum_{j=1}^{m} \int_{\Gamma\times]0,T[} | C_j\varphi |^2 \, dt \right)^{\frac{1}{2}} = \|\{\varphi^0, \varphi^1\}\|_F \quad (13.79)$$

defines a norm on the couple $\{\varphi^0, \varphi^1\}$.

F is therefore the space of completion of smooth functions for the norm (13.79), and is subsequently a Hilbert space with the inner product given by $\langle \Lambda\{\tilde\varphi^0, \tilde\varphi^1\}, \{\varphi^0, \varphi^1\}\rangle$. ∎

The important point is that (13.79) is a norm that corresponds exactly to having the following uniqueness theorem: If

$$\begin{cases} \varphi'' + A\varphi = 0, \\ B_j\varphi = 0 \; ; \; 1 \leq j \leq m, \\ C_j\varphi = 0 \; ; \; 1 \leq j \leq m \end{cases}$$

and T large enough, then $\varphi = 0$. ☐

LEMMA 13.9
Once F is defined as the space of completion of smooth functions for the norm (13.79), Λ is by construction an isomorphism between F and F' where F' denotes F's dual space.

PROOF
Since $\langle \Lambda\{\tilde\varphi^0, \tilde\varphi^1\}, \{\varphi^0, \varphi^1\}\rangle$ defines a scalar product on the set of initial data and our norm is defined by this scalar product, F is a Hilbert space. We

know from Riesz' Representation Theorem that there exists an isomorphic mapping K between F and its dual F' defined by

$$\langle u, v \rangle_F = \langle Ku, v \rangle_{F', F}. \tag{13.80}$$

Writing this in an another way,

$$\langle \Lambda u, v \rangle = \langle Ku, v \rangle_{F', F}. \tag{13.81}$$

We thereby conclude that

$$\Lambda \text{ is isomorphic mapping of } F \text{ onto } F'. \tag{13.82}$$

∎

We can therefore conclude that:

If the couple $\{y^1, y^0\}$ lie in F', then the method described above will succeed in finding the required controls, i.e., (13.68) is achieved.

REMARK 13.7   Let us assume we have a uniqueness theorem of the following type: If

$$\begin{cases} \varphi'' + A\varphi = 0, \\ B_j\varphi = 0 \text{ on } \Gamma \times]0, T[, \ 1 \le j \le m, \\ C_j\varphi = 0 \text{ on } \Gamma_0 \times]0, T[, \ \Gamma_0 \subset \Gamma, \\ j \in J \subset [1, ..., m] \end{cases} \tag{13.83}$$

for T large enough, then $\varphi = 0$.

$$\left(\sum_{j \in J} \int_{\Gamma_0 \times]0, T[} | C_j\varphi |^2 \, d\Gamma dt \right)$$

defines a norm and HUM can now be applied.   ∎

To recapitulate, the theoretical basis of HUM is the observation that if one has the uniqueness of solutions of linear evolutionary systems in a Hilbert space , it is possible to introduce a Hilbert space norm $\| \cdot \|_F$ based on the uniqueness property in such a way that the dual system is exactly controllable in the dual space F'.

The problem of exact controllability is thus transferred to identifying and otherwise characterizing the spaces $\{F, F'\}$. This is essentially a problem in partial differential equations which can be formulated as follows: When the evolution of a system is governed by partial differential equations, can *a priori* estimates of $\|\cdot\|_F$ then be obtained in terms of the norms of spaces which are readily identifiable ?

In practice, it is common to first derive an *a priori* estimate leading to the uniqueness result and then use that as the starting point for the application of HUM. Each such estimate will lead to some exact controllability problem. However, the deriving of such *a priori* estimates should not obscure the simple duality principle underlying the method, as well as the fact that the estimates as such are not a part of the *basic principle*, but only the means by which the space F is identified. Of course, at the practical level the identification of F is the crucial point in order to say that the problem of exact controllability has been solved in any real sense.

13.3 The Variable Coefficients Case

In Theorem 13.1 we obtained exact controllability of the wave equation. The purpose of this chapter is to extend this result to second-order hyperbolic systems with variable coefficients and Dirichlet boundary control. We present two theorems which give different estimates of the time T_0 needed to obtain exact controllability. The first Theorem 13.11 is based on *a priori* estimates given in Ho [11]. Later, Komornik, in [14], presented a simplification of this estimate on which Theorem 13.12 is based. Moreover, the estimate extends to a more general class of operators which contain those assumed by Ho [11].

The purpose of the following section is to state some preliminary lemmas and results.

13.3.1 Notation and Properties of the Operator A

In this section we list the properties of the differential operator A. We assume that

$$a_{ij}, \ a'_{ij} \in L^\infty(\Omega), \qquad \frac{\partial}{\partial x_k}(a_{ij}) \in L^\infty(\Omega), \qquad a_{ij} = a_{ji}$$

and that there exists positive constants α and β such that

$$\alpha|\xi|^2 \le a_{ij}(x)\xi_i\xi_j \le \beta|\xi|^2 \ \text{ for all } x \in \Omega, \xi \in R^n. \tag{13.84}$$

By setting $m_k = (x_k - x_k^0)$, it follows from the above that if we fix the point x^0 such that the differential operator $\frac{\partial}{\partial x_i}(2a_{ij} - m_k \frac{\partial a_{ij}}{\partial x_k}\frac{\partial}{\partial x_j})$ is elliptic, then there exists a positive constant γ such that

$$(2a_{ij} - m_k \frac{\partial a_{ij}}{\partial x_k}) \xi_i \xi_j \geq \gamma a_{ij} \xi_i \xi_j \quad \text{for all } x \in \Omega, \xi \in R^n. \tag{13.85}$$

Let η denote the smallest positive number such that

$$(m_k \xi_k)^2 \leq \eta^2 a_{ij} \xi_i \xi_j \quad \text{for all } x \in \Omega, \xi \in R^n \tag{13.86}$$

and let τ be defined by

$$\tau = \sup_{x \in \Omega} \left(\sum_{i,j,k} \left| \frac{\partial}{\partial x_k} a_{ij} \right|^2 \right)^{\frac{1}{2}} \tag{13.87}$$

and R by

$$R = \sup_{x \in \Omega} |x - x^0|. \tag{13.88}$$

Finally, let λ_0 be such that

$$\|\varphi\|_{L^2(\Omega)}^2 \leq \lambda_0 \|\varphi\|_{H_0^1(\Omega)}, \quad \forall \varphi \in H_0^1(\Omega).$$

13.3.2   Preliminary Lemmas

LEMMA 13.10
Let $\varphi \in C([0,T]; H^2(\Omega)) \cap C^1([0,T]; H^1(\Omega)) \cap C^2([0,T]; L^2(\Omega))$ be an arbitrary function satisfying

$$\varphi'' - A\varphi = 0 \quad \text{in } \Omega \times]0, T[\tag{13.89}$$

and let C be an arbitrary real number. Then

$$\int_{\Gamma \times]0,T[} \nu_i a_{ij} \frac{\partial \varphi}{\partial x_j} \left(2m_k \frac{\partial \varphi}{\partial x_k} + C\varphi \right) + m_k \nu_k \left((\varphi')^2 - a_{ij} \frac{\partial \varphi}{\partial x_i} \frac{\partial \varphi}{\partial x_j} \right) d\Gamma dt$$

$$= \int_{\Omega \times]0,T[} (n - C)(\varphi')^2 + \left((C - n + 2)a_{ij} - m_k \frac{\partial a_{ij}}{\partial x_k} \right) \frac{\partial \varphi}{\partial x_i} \frac{\partial \varphi}{\partial x_j} dx dt$$

$$+ \left[\int_\Omega \varphi'(2m_k \frac{\partial \varphi}{\partial x_k} + C\varphi) dx \right]_0^T. \tag{13.90}$$

PROOF
   We integrate

$$\int_{\Omega \times]0,T[} (2m_k \frac{\partial \varphi}{\partial x_k}) \left(\varphi'' - \frac{\partial}{\partial x_i} \left(a_{ij} \frac{\partial \varphi}{\partial x_j} \right) \right) dx dt = 0$$

by parts to obtain (13.90). For the first term we have

$$\int_{\Omega\times]0,T[} 2m_k \frac{\partial\varphi}{\partial x_k}\varphi''\,dx dt = \int_\Omega 2\varphi' m_k \frac{\partial\varphi}{\partial x_k}|_0^T$$

$$-\int_{\Omega\times]0,T[}(2m_k\frac{\partial\varphi'}{\partial x_k})\varphi'\,dxdt = \left[\int_\Omega 2\varphi' m_k \frac{\partial\varphi}{\partial x_k}dx\right]_0^T$$

$$-\int_{\Omega\times]0,T[} m_k \frac{\partial\varphi'}{\partial x_k}dxdt = \left[\int_\Omega 2\varphi' m_k \frac{\partial\varphi}{\partial x_k}dx\right]_0^T$$

$$-\int_{\Gamma\times]0,T[} m_k\nu_k(\varphi')^2 d\Gamma dt + \int_{\Omega\times]0,T[} n(\varphi')^2 dxdt$$

and if we use the symmetry of a_{ij} we have for the second term that

$$\int_{\Omega\times]0,T[} 2m_k \frac{\partial\varphi}{\partial x_k}\frac{\partial}{\partial x_i}\left(a_{ij}\frac{\partial\varphi}{\partial x_j}\right) dxdt \qquad (13.91)$$

$$= \int_{\Gamma\times]0,T[} \nu_i a_{ij}\frac{\partial\varphi}{\partial x_j} 2m_k \frac{\partial\varphi}{\partial x_k} d\Gamma dt - \int_{\Omega\times]0,T[} a_{ij}\frac{\partial\varphi}{\partial x_j}\frac{\partial}{\partial x_i}\left(2m_k\frac{\partial\varphi}{\partial x_k}\right)dxdt$$

$$= \int_{\Gamma\times]0,T[} \nu_i a_{ij}\frac{\partial\varphi}{\partial x_j} 2m_k \frac{\partial\varphi}{\partial x_k} d\Gamma dt - \int_{\Omega\times]0,T[} 2\frac{\partial h_k}{\partial x_i}a_{ij}\frac{\partial\varphi}{\partial x_j}\frac{\partial\varphi}{\partial x_k}dxdt$$

$$-\int_{\Omega\times]0,T[} m_k \frac{\partial}{\partial x_k}\left(a_{ij}\frac{\partial\varphi}{\partial x_i}\right)\frac{\partial\varphi}{\partial x_j}dxdt + \int_{\Omega\times]0,T[} m_k \frac{\partial a_{ij}}{\partial x_k}\frac{\partial\varphi}{\partial x_i}\frac{\partial\varphi}{\partial x_j}dxdt$$

$$= \int_{\Gamma\times]0,T[} \nu_i a_{ij}\frac{\partial\varphi}{\partial x_j} 2m_k \frac{\partial\varphi}{\partial x_k} d\Gamma dt - \int_{\Omega\times]0,T[} 2a_{ij}\frac{\partial\varphi}{\partial x_i}\frac{\partial\varphi}{\partial x_j}dxdt$$

$$-\int_{\Gamma\times]0,T[} m_k\nu_k a_{ij}\frac{\partial\varphi}{\partial x_i}\frac{\partial\varphi}{\partial x_j}d\Gamma dt + \int_{\Omega\times]0,T[} n a_{ij}\frac{\partial\varphi}{\partial x_i}\frac{\partial\varphi}{\partial x_j}dxdt$$

$$+\int_{\Omega\times]0,T[} m_k \frac{\partial a_{ij}}{\partial x_k}\frac{\partial\varphi}{\partial x_i}\frac{\partial\varphi}{\partial x_j}dxdt.$$

Hence, the case $C = 0$ of the lemma follows. To conclude in the general case, it is sufficient to prove that

$$\int_{\Gamma\times]0,T[} \varphi\nu_i a_{ij}\frac{\partial\varphi}{\partial x_j} d\Gamma dt = \int_{\Omega\times]0,T[} a_{ij}\frac{\partial\varphi}{\partial x_i}\frac{\partial\varphi}{\partial x_j} - (\varphi')^2 dxdt + \left[\int_\Omega \varphi\varphi' dx\right]_0^T.$$

This follows by integrating by parts in the identity

$$\int_{\Omega\times]0,T[} \varphi\left(\varphi'' - \frac{\partial}{\partial x_i}\left(a_{ij}\frac{\partial\varphi}{\partial x_j}\right)\right) dxdt = 0$$

to obtain

$$\int_{\Omega\times]0,T[} \varphi\varphi'' dxdt = \left[\int_\Omega \varphi\varphi' dx\right]_0^T - \int_{\Omega\times]0,T[}(\varphi')^2 dxdt$$

and

$$\int_{\Omega\times]0,T[} \varphi \frac{\partial}{\partial x_i}\left(a_{ij}\frac{\partial\varphi}{\partial x_j}\right)dxdt = \int_{\Gamma\times]0,T[}\varphi\nu_i a_{ij}\frac{\partial\varphi}{\partial x_j}d\Gamma dt - \int_{\Omega\times]0,T[}a_{ij}\frac{\partial\varphi}{\partial x_i}\frac{\partial\varphi}{\partial x_j}dx$$

We integrate

$$\int_\Omega \frac{1}{2}m_k \frac{\partial\varphi}{\partial x_k}\left(\varphi'' - \frac{\partial}{\partial x_i}\left(a_{ij}\frac{\partial\varphi}{\partial x_j}\right)\right)dx = 0$$

by parts to obtain the result. For the first term we have

$$\int_\Omega \frac{1}{2}m_k\frac{\partial\varphi}{\partial x_k}(\varphi'')dx = \int_\Omega \frac{1}{2}m_k\frac{\partial\varphi}{\partial x_k}(\varphi'')dx$$
$$+ \int_\Omega \frac{1}{2}m_k\frac{\partial\varphi'}{\partial x_k}\frac{\partial\varphi}{\partial t}dx - \int_\Omega \frac{1}{2}m_k\frac{\partial\varphi'}{\partial x_k}\frac{\partial\varphi}{\partial t}dx$$
$$= n\frac{1}{2}\int_\Omega \left(\frac{\partial\varphi}{\partial t}\right)^2 dx + \frac{\partial}{\partial t}\int_\Omega m_k\frac{\partial\varphi}{\partial x_k}\frac{\partial\varphi}{\partial t}dx$$

and, since $\frac{\partial\varphi}{\partial x_k} = \frac{\partial\varphi}{\partial\nu}\nu_k$ on Γ, we have from (13.91), for the second term, that

$$\int_\Omega \frac{1}{2}m_k\frac{\partial\varphi}{\partial x_k}\frac{\partial}{\partial x_i}\left(a_{ij}\frac{\partial\varphi}{\partial x_j}\right)dx$$
$$= \int_\Gamma \nu_i a_{ij}\frac{\partial\varphi}{\partial x_j}m_k\frac{\partial\varphi}{\partial x_k}d\Gamma dt - \int_\Omega a_{ij}\frac{\partial\varphi}{\partial x_i}\frac{\partial\varphi}{\partial x_j}dx$$
$$- \frac{1}{2}\int_{\Gamma\times]0,T[} m_k\nu_k a_{ij}\frac{\partial\varphi}{\partial x_i}\frac{\partial\varphi}{\partial x_j}d\Gamma dt + \int_\Omega n\frac{1}{2}a_{ij}\frac{\partial\varphi}{\partial x_i}\frac{\partial\varphi}{\partial x_j}dx$$
$$+ \int_\Omega \frac{1}{2}m_k\frac{\partial a_{ij}}{\partial x_k}\frac{\partial\varphi}{\partial x_i}\frac{\partial\varphi}{\partial x_j}dx$$
$$= (n-2)\int_\Omega \frac{1}{2}a_{ij}\frac{\partial\varphi}{\partial x_i}\frac{\partial\varphi}{\partial x_j}dx + \int_{\Gamma\times]0,T[}\frac{1}{2}m_k\nu_k a_{ij}\frac{\partial\varphi}{\partial x_i}\frac{\partial\varphi}{\partial x_j}d\Gamma dt$$
$$+ \int_\Omega \frac{1}{2}m_k\frac{\partial a_{ij}}{\partial x_k}\frac{\partial\varphi}{\partial x_i}\frac{\partial\varphi}{\partial x_j}dx$$

and we obtain the result. ∎

13.3.3   Application of HUM

Using the results of the previous section, we are able to extend the result of Theorem 13.1 to second-order hyperbolic systems with variable coefficients.

THEOREM 13.11

Consider $\Gamma_0 \subset \Gamma$ defined by Definition 13.1. If

$$\sup_{x \in \Omega} \left(\sum_{i,j,k} \left| \frac{\partial}{\partial x_k} a_{ij} \right|^2 \right)^{\frac{1}{2}} < \frac{2\alpha}{\sup_{x \in \Omega} m_k},$$

then for any initial conditions $\{y^0, y^1\} \in L^2(\Omega) \times H^{-1}(\Omega)$, one can find a control $\kappa \in L^2(\Gamma_0 \times]0, T[)$ such that κ drives the system

$$\begin{cases} y'' - Ay = 0 \text{ in } \Omega \times]0, T[, \\ y = \kappa \text{ on } \Gamma_0 \times]0, T[, \\ y = 0 \text{ on } \Gamma \backslash \Gamma_0, \\ y(0) = y^0, y'(0) = y^1 \text{ on } \Omega \end{cases}$$

to rest in finite time $T > T_0$, where

$$T_0 = \left(\frac{2R}{\alpha} + \frac{1}{\lambda_0} \left(n - 1 + \frac{R\tau}{2\alpha} \right) \right) \frac{2\alpha}{2\alpha - R\tau}.$$

If we use an estimate due to Komornik [14], we have the Theorem 13.12 below.

THEOREM 13.12

Consider $\Gamma_0 \subset \Gamma$ defined by Definition 13.1. If x^0 is fixed such that the differential operator

$$\frac{\partial}{\partial x_i} \left(2a_{ij} - m_k \frac{\partial a_{ij}}{\partial x_k} \frac{\partial}{\partial x_j} \right) \tag{13.92}$$

is elliptic, then for any initial conditions $\{y^0, y^1\} \in L^2(\Omega) \times H^{-1}(\Omega)$, one can find a control $\kappa \in L^2(\Gamma_0 \times]0, T[)$ such that κ drives the system

$$\begin{cases} y'' - Ay = 0 \text{ in } \Omega \times]0, T[, \\ y = \kappa \text{ on } \Gamma \times]0, T[, \\ y = 0 \text{ on } \Gamma \backslash \Gamma_0 \times]0, T[\\ y(0) = y^0, y'(0) = y^1 \text{ on } \Omega \end{cases}$$

to rest in finite time $T > T_0 = \frac{4\eta}{\min\{\gamma, 2n\}}$.

The estimate of T_0 given by Theorem 13.12 is simpler than the one of Theorem 13.11. Moreover, Theorem 13.12 can be applied to a larger class of operators than in Theorem 13.11. The condition on the choice of x^0 in

Theorem 13.12 is that we have to fix x_0 such that the differential operator $\frac{\partial}{\partial x_i}(2a_{ij} - m_k \frac{\partial a_{ij}}{\partial x_k} \frac{\partial}{\partial x_j})$ is elliptic. This implies (cf. (13.85)) that

$$(2a_{ij} - m_k \frac{\partial a_{ij}}{\partial x_k})\xi_i \xi_j \geq \gamma a_{ij}\xi_i \xi_j \text{ for all } x \in \Omega, \xi \in R^n. \tag{13.93}$$

It is easily seen that the assumption on the coefficients in Theorem 13.11

$$\sup_{x \in \Omega} \left(\sum_{i,j,k} \left| \frac{\partial}{\partial x_k} a_{ij} \right|^2 \right)^{\frac{1}{2}} < \frac{2\alpha}{\sup_{x \in \Omega} m_k}$$

implies (13.93). It is also seen that in both cases, x^0 cannot be chosen arbitrarily in R^n, but if we fix x^0 we can characterize those systems we are able to control.

The estimate of T_0 in Theorem 13.12 is optimal in the sense that there does not exist an estimate $\tilde{T}_0 < T_0$ (cf. Lagnese [15] and Bardos-Lebeau-Rauch [4]). Unfortunately, we cannot give an estimate of T_0 which only depends on the geometry of Ω as in Theorem 13.1.

REMARK 13.8   It still remains an open question whether we can obtain the results of Theorem 13.11 and Theorem 13.12 with an arbitrary x^0.   ∎

THE PROOF OF THEOREM 13.11:

Following the idea of HUM, we consider two systems, the φ-system:

$$\begin{cases} \varphi'' - A\varphi = 0 \text{ in } \Omega \times]0, T[, \\ \varphi = 0 \text{ on } \Gamma \times]0, T[, \\ \varphi(0) = \varphi^0, \varphi'(0) = \varphi^1 \text{ in } \Omega. \end{cases}$$

and the ψ-system:

$$\psi'' - A\psi = 0 \text{ in } \Omega \times]0, T[,$$

$$\psi = \begin{cases} \frac{\partial \varphi}{\partial \nu} \text{ on } \Gamma_0 \times]0, T[, \\ 0 \text{ on } \Gamma \backslash \Gamma_0 \times]0, T[\end{cases}$$

with

$$\psi(x, T) = \psi'(x, T) = 0 \text{ in } \Omega.$$

The existence and regularity of the solutions of the systems above are given in Section 12.5, Chapter 12.

From (13.75) in Section 13.2.2, we can derive that

$$\langle\Lambda\{\varphi^0,\varphi^1\},\{\varphi^0,\varphi^1\}\rangle = \langle\{\psi'(0),-\psi(0)\},\{\varphi^0,\varphi^1\}\rangle = \int_{\Gamma_0\times]0,T[}\left|\frac{\partial\varphi}{\partial\nu}\right|^2 d\Gamma dt.$$

(13.94)

The main point is then to show that

$$\langle\Lambda\{\varphi^0,\varphi^1\},\{\varphi^0,\varphi^1\}\rangle = \int_{\Gamma_0\times]0,T[}\left|\frac{\partial\varphi}{\partial\nu}\right|^2 d\Gamma dt \qquad (13.95)$$

defines a norm for T large enough. If we follow the lines of the proof of Theorem 13.1, we only we have show that there exist constants c_1 and $c_2(T)$ such that

$$\int_{\Gamma_0\times]0,T[}\left|\frac{\partial\varphi}{\partial\nu}\right|^2 d\Gamma dt \geq K(T-T_0)E_0 \qquad (13.96)$$

$$\geq c_1\left[\|\varphi^0\|^2_{H^1_0(\Omega)} + \|\varphi^1\|^2_{L^2(\Omega)}\right], \qquad (13.97)$$

$$\int_{\Gamma_0\times]0,T[}\left|\frac{\partial\varphi}{\partial\nu}\right|^2 d\Gamma dt \leq c_2(T)\left[\|\varphi^0\|^2_{H^1_0(\Omega)} + \|\varphi^1\|^2_{L^2(\Omega)}\right], \qquad (13.98)$$

where the energy E is defined by

$$E_0 = E = \int_\Omega\left(\frac{\partial\varphi}{\partial t}\right)^2 dx + \int_\Omega a_{ij}\frac{\partial\varphi}{\partial x_i}\frac{\partial\varphi}{\partial x_j}dx$$

and is independent of t. The reason why we only have to show this is that if we show (13.96), Λ is clearly a self-adjoint and positive operator for $T > T_0$. Moreover, from (13.96)-(13.98) this norm is equivalent to the norm of $H^1_0(\Omega) \times L^2(\Omega)$. Then we can derive that Λ is an isomorphic mapping of $H^1_0(\Omega) \times L^2(\Omega)$ onto $H^{-1}(\Omega) \times L^2(\Omega)$.

We will first show (13.96).

STEP 1. *An* a priori *estimate*

From Lemma 13.9 we have the identity

$$n\int_\Omega\frac{1}{2}\left(\frac{\partial\varphi}{\partial t}\right)^2 dx - (n-2)\int_{\Omega\times]0,T[}a_{ij}\frac{\partial\varphi}{\partial x_i}\frac{\partial\varphi}{\partial x_j}dxdt$$

$$+ \frac{\partial}{\partial t}\int_\Omega m_k\frac{\partial\varphi}{\partial x_k}\frac{\partial\varphi}{\partial t}dx$$

$$- \int_{\Gamma\times]0,T[}\frac{1}{2}m_k\nu_k a_{ij}\nu_i\nu_j\left(\frac{\partial\varphi}{\partial\nu}\right)d\Gamma dt$$

$$- \int_{\Omega\times]0,T[}\frac{1}{2}m_k\frac{\partial a_{ij}}{\partial x_k}\frac{\partial\varphi}{\partial x_i}\frac{\partial\varphi}{\partial x_j}dxdt = 0. \qquad (13.99)$$

Hence, (13.84), (13.87), (13.88), and (13.99) imply

$$n \int_\Omega \frac{1}{2} \left(\frac{\partial\varphi}{\partial t}\right)^2 dx - (n - 2 + \frac{R\tau}{2\alpha}) \int_\Omega a_{ij} \frac{\partial\varphi}{\partial x_i} \frac{\partial\varphi}{\partial x_j} dx \quad (13.100)$$
$$+ \frac{\partial}{\partial t} \int_\Omega m_k \frac{\partial\varphi}{\partial x_k} \frac{\partial\varphi}{\partial t} dx - \frac{1}{2}\gamma R \int_{\Gamma_0} \left(\frac{\partial\varphi}{\partial\nu}\right) d\Gamma \le 0.$$

STEP 2. A priori *estimate*
If we integrate by parts the identity

$$\frac{1}{2} \int_\Omega u(\varphi'' - \frac{\partial}{\partial x_i}(a_{ij}\frac{\partial\varphi}{\partial x_j}))dx = 0,$$

we obtain

$$\int_\Omega \frac{1}{2} \left(\frac{\partial\varphi}{\partial t}\right)^2 dx - \int_{\Omega\times]0,T[} a_{ij} \frac{\partial\varphi}{\partial x_i} \frac{\partial\varphi}{\partial x_j} dx dt - \frac{1}{2}\frac{\partial}{\partial t} \int_\Omega \varphi \frac{\partial\varphi}{\partial t} dx = 0.$$
$$(13.101)$$

Hence, if we multiply (13.101) by $-(n - 1 + \frac{R\tau}{2\alpha})$ and add it to (13.100), we immediately obtain

$$\left(1 - \frac{R\alpha}{2\tau}\right) E_0\varphi \le \frac{1}{2}\gamma R \int_{\Gamma_0} \left(\frac{\partial\varphi}{\partial\nu}\right)^2 d\Gamma \quad (13.102)$$
$$- \frac{\partial}{\partial t} \int_\Omega \left(m_k \frac{\partial\varphi}{\partial x_k} + (\frac{n - 1}{2} + \frac{R\tau}{4\alpha})\varphi\right) \frac{\partial\varphi}{\partial t} dx.$$

Integrating (13.102) from $t = 0$ to $t = T$, we obtain

$$T\left(1 - \frac{R\alpha}{2\tau}\right) E_0\varphi \le \frac{1}{2}\gamma R \int_0^T \int_{\Gamma_0} \left(\frac{\partial\varphi}{\partial\nu}\right)^2 d\Gamma dt \quad (13.103)$$
$$- \left[\int_\Omega \left(m_k \frac{\partial\varphi}{\partial x_k} + (\frac{n - 1}{2} + \frac{R\tau}{4\alpha})\varphi\right) \frac{\partial\varphi}{\partial t} dx\right]_{t=0}^{t=T}.$$

STEP 3. *The estimate (13.96)*
From (13.103) we have that

$$T\left(1 - \frac{R\alpha}{2\tau}\right) E_0\varphi \le \frac{1}{2}\gamma R \int_0^T \int_{\Gamma_0} \left(\frac{\partial\varphi}{\partial\nu}\right)^2 d\Gamma dt$$
$$+ R\left(\|\nabla\varphi(x,T)\|_{L^2(\Omega)}\|\varphi'(x,T)\|_{L^2(\Omega)} + \|\nabla\varphi(x,0)\|_{L^2(\Omega)}\|\varphi'(x,0)\|_{L^2(\Omega)}\right)$$
$$+ \left(\frac{n - 1}{2} + \frac{R\tau}{4\alpha}\right)\left(\|\nabla\varphi(x,T)\|_{L^2(\Omega)}\|\varphi'(x,T)\|_{L^2(\Omega)}\right)$$
$$+ \|\nabla\varphi(x,0)\|_{L^2(\Omega)}\|\varphi'(x,0)\|_{L^2(\Omega)}).$$

If we recall that $|ab| \leq a^2 + b^2$ for $a, b \in R$ and use the facts that

$$\|\nabla\varphi\|^2_{L^2(\Omega)} \leq \frac{2}{\alpha} \int_\Omega a_{ij} \frac{\partial\varphi}{\partial x_i} \frac{\partial\varphi}{\partial x_j} dx \qquad (13.104)$$

and, since $\|\varphi\|^2_{L^2(\Omega)} \leq \lambda_0 \|\nabla\varphi\|^2_{L^2(\Omega)}$ for all $\varphi \in H^1_0(\Omega)$,

$$\|\varphi\|^2_{L^2(\Omega)} \leq \frac{2}{\lambda_0} \int_\Omega a_{ij} \frac{\partial\varphi}{\partial x_i} \frac{\partial\varphi}{\partial x_j} dx \qquad (13.105)$$

for all t we have

$$T\left(1 - \frac{R\alpha}{2\tau}\right) E_0\varphi \leq \frac{1}{2}\gamma R \int_{\Gamma_0 \times]0,T[} \left(\frac{\partial\varphi}{\partial\nu}\right)^2 d\Gamma dt$$
$$+ \left(\frac{2R}{\alpha} + \frac{1}{\lambda_0}(n-1) + \frac{R\tau}{2\alpha}\right) E_0\varphi. \qquad (13.106)$$

Since $\tau < \frac{2\alpha}{R}$, we may divide (13.106) by $1 - \frac{R\tau}{2\alpha}$ and so obtain the desired inequality (13.96).

STEP 4. *The estimate (13.98)*
The inequality (13.98) follows directly by setting $F = g = 0$ in Theorem 12.15, Chapter 12. This completes the proof of Theorem 13.10.

THE PROOF OF THEOREM 13.12.

As in the proof of Theorem 13.11, we only have to show that there exists a constant c_1 such that

$$\int_{\Gamma_0 \times]0,T[} \left|\frac{\partial\varphi}{\partial\nu}\right|^2 d\Gamma dt \geq (T - T_0)E_0 \geq c_1 \left[\|\varphi^0\|^2_{H^1_0(\Omega)} + \|\varphi^1\|^2_{L^2(\Omega)}\right]$$
$$(13.107)$$

and a constant $c_2(T)$ such that

$$\int_{\Gamma_0 \times]0,T[} \left|\frac{\partial\varphi}{\partial\nu}\right|^2 d\Gamma dt \leq c_2(T) \left[\|\varphi^0\|^2_{H^1_0(\Omega)} + \|\varphi^1\|^2_{L^2(\Omega)}\right]. \qquad (13.108)$$

We will first show (13.107). From Lemma 13.10 we have for any constant

C that

$$\underbrace{\int_{\Gamma\times]0,T[} \nu^i a_{ij}\frac{\partial\varphi}{\partial x_j}\left(2m_k\frac{\partial\varphi}{\partial x_k}+C\varphi\right)+m_k\nu_k\left((\varphi')^2-a_{ij}\frac{\partial\varphi}{\partial x_i}\frac{\partial\varphi}{\partial x_j}\right)d\Gamma dt}_{I}$$

$$=\underbrace{\int_{\Omega\times]0,T[}(n-C)(\varphi')^2+\left((C-n+2)a_{ij}-m_k\frac{\partial a_{ij}}{\partial x_k}\right)\frac{\partial\varphi}{\partial x_i}\frac{\partial\varphi}{\partial x_j}dxdt}_{II}+$$

$$+\underbrace{\left[\int_\Omega\varphi'(2m_k\frac{\partial\varphi}{\partial x_k}+C\varphi)dx\right]_0^T}_{III}.$$

STEP 1. A priori *estimate I*
On $\Gamma\times]0,T[$ we have $\frac{\partial\varphi}{\partial x_j}=\frac{\partial\varphi}{\partial\nu}\nu_j$ and $\varphi=0$. Hence

$$\nu^i\nu^j a_{ij}\frac{\partial\varphi}{\partial\nu}(2m_k\frac{\partial\varphi}{\partial x_k}+C\varphi)+m_k\nu_k((\varphi')^2-a_{ij}\frac{\partial\varphi}{\partial x_i}\frac{\partial\varphi}{\partial x_j}) \quad (13.109)$$

$$=\nu^i\nu^j a_{ij}m_k\nu_k\left(\frac{\partial\varphi}{\partial\nu}\right)^2.$$

We conclude due to (13.84) that

$$I\le\int_{\Gamma_0\times]0,T[}\nu^i\nu^j a_{ij}m_k\nu_k\left(\frac{\partial\varphi}{\partial\nu}\right)^2 d\Gamma dt. \quad (13.110)$$

STEP 2. A priori *estimate II*
From (13.85) we have that $2a_{ij}-m_k\frac{\partial a_{ij}}{\partial x_k}\xi_i\xi_j\ge\gamma a_{ij}\xi_i\xi_j$, and using this in
II we get

$$II\ge\int_{\Omega\times]0,T[}(n-C)(\varphi')^2+\left(2(C-n)\frac{a_{ij}}{2}-2\gamma\frac{a_{ij}}{2}\right)\frac{\partial\varphi}{\partial x_i}\frac{\partial\varphi}{\partial x_j}dxdt. \quad (13.111)$$

If we choose $C=\max\{n-\frac{\gamma}{2},0\}$ we obtain that

$$II\ge\int_{\Omega\times]0,T}2(n-\max\{n-\frac{\gamma}{2},0\})\left[\frac{(\varphi')^2}{2}+\frac{a_{ij}}{2}\frac{\partial\varphi}{\partial x_i}\frac{\partial\varphi}{\partial x_j}\right]dxdt \quad (13.112)$$

$$=2(n-\max\{n-\frac{\gamma}{2},0\})E_0, \quad (13.113)$$

since the energy E is constant.
STEP 3 A priori *estimate III*

From Lemma 13.10 we have

$$\left| [\int_\Omega \varphi'(2m_k\frac{\partial\varphi}{\partial x_k} + C\varphi)dx]_0^T \right| \leq 2\eta E_0 - C(2n - C)(4\eta)^{-1}\int_\Omega (\varphi)^2 dx$$
$$+ C(2\eta)^{-1}\int_{\Gamma_0} m_k\nu_k\varphi^2 d\Gamma.$$

The first term on the R.H.S. is zero, and the second term is always less than or equal to zero by the choice of C. Hence,

$$\left| [\int_\Omega \varphi'(2m_k\frac{\partial\varphi}{\partial x_k} + C\varphi)dx]_0^T \right| \leq 2\eta E_0. \qquad (13.114)$$

STEP 4 *The inequality (13.107)*
Combining (13.110), (13.112), and (13.114), we obtain

$$\int_{\Gamma_0\times]0,T[} a_{ij}\nu_i\nu_j m_k\nu_k \left(\frac{\partial\varphi}{\partial\nu}\right)^2 d\Gamma dt \geq \min\{\gamma, 2n\}(T - T_0)E_0, \qquad (13.115)$$

where $T_0 = \frac{4\eta}{\min\{\gamma,2n\}}$. Since $a_{ij} \in L^\infty(\Omega)$, there exists a constant k such that

$$\int_{\Gamma_0\times]0,T[} \left(\frac{\partial\varphi}{\partial\nu}\right)^2 d\Gamma dt \geq \frac{\min\{\gamma, 2n\}}{k}(T - T_0)E_0. \qquad (13.116)$$

STEP 5 $E_0 \geq \left[\|\varphi^0\|_{H_0^1(\Omega)}^2 + \|\varphi^1\|_{L^2(\Omega)}^2\right]$
Using (13.84) we have

$$\int_{\Gamma_0\times]0,T[} a_{ij}\frac{\partial\varphi}{\partial x_i}\frac{\partial\varphi}{\partial x_j} \geq \frac{\alpha}{2}\|\nabla\varphi(x, 0)\|_{L^2(\Omega)}^2 \qquad (13.117)$$

and since

$$E_0 = \int_\Omega \frac{1}{2}|\varphi'(0)|^2 + a_{ij}\frac{\partial\varphi(0)}{\partial x_i}\frac{\partial\varphi(0)}{\partial x_j}dx,$$

we immediately have (13.107).
STEP 6. *The inequality (13.108)*
The inequality (13.108) follows directly by setting $F = g = 0$ in Theorem 12.15, Chapter 12. This completes the proof.

Exercises

Exercise 1
Let φ be a real continuous function defined for $x \geq 0$, and assume that $lim_{x \to \infty}\varphi(x)$ exists (and is finite). Show that for $\epsilon > 0$ there are $n \in N$ and constants $a_k, k = 0, 1, ..., n$, such that

$$|\varphi(x) - \sum_{k=0}^{n} a_k e^{-kx}| \leq \epsilon$$

for all $x \geq 0$. ☐

Exercise 2
Let (M, d) be a metric space.
We define the *open ball* with center x_0 and radius $r > 0$ by

$$B(x_0, r) = \{x \in M \mid d(x, x_0) < r\}.$$

We denote a subset $A \subset M$ *open*, if for any $x_0 \in A$, there is an open ball with center x_0 contained in A.
Show that an open ball is an open set.
Show that the set of open sets defined in this way is a topology on M. ☐

Exercise 3
Let (M, d) be a metric space. We say that a mapping $T : M \to M$ is *continuous* in $x_0 \in M$ if for any $\epsilon > 0$ there is a $\delta > 0$ such that for all $x \in M$ we have
$$d(x_0, x) < \delta \Rightarrow d(Tx_0, Tx) < \epsilon.$$

Show that T is continuous in x_0 if and only if

$$x_n \to x_0 \Rightarrow Tx_n \to Tx_0.$$

Show that T is continuous in the sense of definition 1.1 if the open sets are defined as in Exercise 2.   ▯

Exercise 4
In a set, M is given a function d' from $M \times M$ to R that satisfies

$d'(x,y) = 0$ if and only if $x = y$

d' $(x,y) \le d'(z,x) + d'(z,y)$ for all $x, y, z \in M$

Show that (M, d') is a metric space.   ▯

Exercise 5
Let (M, d) be a metric space.
The *diameter* of a nonempty subset A of M is defined as

$$\delta(A) = \sup_{x,y \in A} d(x,y) \quad (\le \infty).$$

Show that $\delta(A) = 0$ if and only if A contains only one point.   ▯

Exercise 6
Let (M, d) be a metric space. Show that d_1 given by

$$d_1(x,y) = \frac{d(x,y)}{1 + d(x,y)} \quad \text{for} \quad x, y \in M$$

is a metric on M.
Show that

$$\delta_1(A) = \sup_{x,y \in A} d_1(x,y) \le 1$$

for all $A \subset M$. Is it possible to find a subset A with $\delta_1(A) = 1$?
Show that $d_1(x_n, x) \to 0$ if and only if $d_0(x_n, x) \to 0$.   ▯

Exercise 7
Let (M_1, d_1) and (M_2, d_2) be metric spaces.
Show that $M_1 \times M_2$ can be made into a metric space by the following definition of a metric d:

$$d((x_1, x_2), (y_1, y_2)) = d_1(x_1, y_1) + d_2(x_2, y_2).$$

Show that d^* given by

$$d^*((x_1, x_2), (y_1, y_2)) = \max\{d_1(x_1, y_1), d_2(x_2, y_2)\}$$

also defines a metric on $M_1 \times M_2$. ⬚

Exercise 8
Show that in any set M we can define a metric by

$$d(x, y) = \begin{cases} 0 \text{ if } x = y \\ 1 \text{ if } x \neq y. \end{cases}$$

Then we call (M, d) for a *discrete* metric space.
Characterize the sequences in M where $d(x_n, x) \to 0$. ⬚

Exercise 9
Let (M, d) be a metric space and consider M as a topological space with the topology stemming from the open balls (the *ball topology*). Recall that a set A is *closed* if $M \setminus A$ is open.
Show that $A \subset M$ is closed if and only if

$$x_n \in A, \quad x_n \to x \Rightarrow x \in A.$$

Show that if (M, d) is a complete metric space and A is a closed subset of M, then (A, d) is a complete metric space. ⬚

Exercise 10
Show that

$$d(x, y) = |\arctan x - \arctan y|$$

defines a metric on R. ⬚

Exercise 11
In R^k we define

$$d_1(x, y) = \sum_{i=1}^{k} |x_i - y_i|,$$

$$d_2(x, y) = (\sum_{i=1}^{k} |x_i - y_i|^2)^{\frac{1}{2}},$$

$$d_\infty = \max_{1 \leq i \leq k} |x_i - y_i|.$$

Show that d_1, d_2 and d_∞ are metrics.
Show that

$$d_\infty(x, y) \leq d_1(x, y) \leq k d_\infty(x, y),$$

and find a similar inequality when d_1 is replaced by d_2.

Show that if a sequence (x_n) converges to x in one of the metrics, then we have coordinatewise convergence:

$$x_{ni} \to x_i$$

for all $i = 1, 2, ..., k$.   ▯

Exercise 12

Let c denote the set of convergent complex sequences $x = (x_1, x_2, ...)$. Show that c is a complete metric space when equipped with the metric

$$d_\infty(x, y) = \sup_i |x_i - y_i|.$$

Hint: Show that the bounded complex sequences l^∞ is a complete space, show that c is a closed subset, and then apply Exercise 9.   ▯

Exercise 13

In the set of bounded complex sequences l^∞ equipped with the metric from Exercise 12, we consider the sets c_0 consisting of the sequences converging to 0 and c_{00} consisting of the sequences with only a finite number of elements different from 0.

Investigate if c_0 and/or c_{00} are closed subsets of l^∞.   ▯

Exercise 14

Consider the metric space (M, d) where $M = [1; \infty)$ and d the usual distance. Let the mapping $T : M \to M$ be given by

$$Tx = \frac{x}{2} + \frac{1}{x}.$$

Show that T is a contraction and find the minimal contraction constant α. Find also the fixed point.   ▯

Exercise 15

A mapping T from a metric space (M, d) into itself is called a *weak contraction* if

$$d(Tx, Ty) < d(x, y),$$

for all $x, y \in M$, $x \neq y$.

Show that T has at most one fixed point.

Show that T does not necessarily have a fixed point. (One could take $Tx = x + \frac{1}{x}$ for $x \geq 1$.)   ▯

Exercise 16
It is very common in mathematical analysis to consider iterations of the form

$$x_n = g(x_{n-1}),$$

where g is a C^1-function. Show that the sequence (x_n) is convergent for any choice of x_0 if there is a α, $0 < \alpha < 1$, such that

$$|g'(x)| \leq \alpha,$$

for all $x \in R$. ☐

Exercise 17
To approximate the solution to an equation $f(x) = 0$, we bring the equation on the form $x = g(x)$, choose an x_0, and use the iteration $x_n = g(x_{n-1})$. Assume that g is a C^1-function on the interval $[x_0 - \delta; x_0 + \delta]$, and that $|g'(x)| \leq \alpha < 1$ for $x \in [x_0 - \delta; x_0 + \delta]$, and moreover

$$|g(x_0) - x_0| \leq (1 - \alpha)\delta.$$

Show that there is one and only one solution $x \in [x_0 - \delta; x_0 + \delta]$ to the equation, and that $x_n \to x$. ☐

Exercise 18
Solve by iteration the equation $f(x) = 0$ for $f \in C^1([a; b])$, $f(a) < 0 < f(b)$ and f' bounded and strictly positive in $[a; b]$. (Take $g(x) = x - \lambda f(x)$ for a smart choice of λ). ☐

Exercise 19
Show that it is possible to solve the equation $f(x) = x^3 + x - 1 = 0$ by the iteration

$$x_n = g(x_{n-1}) = (1 + x_{n-1}^2)^{-1}.$$

Find x_1, x_2, x_3 for $x_0 = 1$, and find an estimate for $d(x, x_n)$. ☐

Exercise 20
A mapping $T : R \to R$ satisfies a *Lipschitz-condition* with constant k if

$$|Tx - Ty| \leq k|x - y|,$$

for all $x, y \in R$.
1: Is T a contraction ?
2: If T is a C^1-function with bounded derivative, show that T satisfies a Lipschitz condition.

3: If T satisfies a Lipschitz condition, is T then a C^1-function with bounded derivative ?

4: Assume that $|Tx - Ty| \leq k|x - y|^\alpha$ for some $\alpha > 1$. Show that T is a constant. ▯

Exercise 21

Let T be a mapping from a complete metric space (M, d) into itself, and assume that there is a natural number m such that T^m is a contraction. Show that T has one and only one fixed point. ▯

Exercise 22

We consider the metric space R^k with the metric $d_1(x, y) = \sum_{i=1}^k |x_i - y_i|$ and a mapping $T : R^k \to R^k$ given by $Tx = Cx + b$, where $C = (c_{ij})$ is a $k \times k$ matrix and $b \in R^k$.

Show that T is a contraction if $\sum_{i=1}^k |c_{ij}| < 1$ for all $j = 1, 2, ..., k$.

If we instead use the metric $d_2(x, y) = \sum_{i=1}^k |x_i - y_i|^2$, show that T is a contraction if

$$\sum_{i=1}^k \sum_{j=1}^k |c_{ij}|^2 < 1.$$

▯

Exercise 23

Consider the *Volterra integral equation*:

$$x(t) - \mu \int_a^t k(t, s)x(s)ds = v(t), \quad t \in [a; b],$$

where $v \in C([a; b])$, $k \in C([a; b]^2)$ and $\mu \in C$.

Show that the equation has a unique solution $x \in C([a; b])$ for any $\mu \in C$.

(Hint: Write the equation $x = Tx$ where

$$Tx = v(t) + \mu \int_a^t k(t, s)x(s)ds.$$

Take $x_0 \in C([a; b])$ and define the iteration by $x_{n+1} = Tx_n$, then show by induction that

$$|T^m x(t) - T^m y(t)| \leq |\mu|^m c^m \frac{(t-a)^m}{m!} d_\infty(x, y),$$

where $c = \max |k|$. Then show (by looking at $d_\infty(T^m x, T^m y)$) that T^m is a contraction for some m and argue that T then must have a unique fixed point in the metric space $(C([a; b]), d_\infty)$. ▯

Exercise 24
Let $f \in L^1(R)$.
1: Can we conclude that $f(x) \to 0$ for $|x| \to \infty$?
2: Can we find $a, b \in R$ such that $|f(x)| \le b$ for $|x| \ge a$?   ☐

Exercise 25
In the vector space $C([a; b])$ we consider the functions $e_0(t), e_1(t), ..., e_n(t)$, where $e_j(t)$ is a polynomial of degree $j, j = 0, 1, ..., n$.
Show that $e_0, e_1, ..., e_n$ are linearly independent.   ☐

Exercise 26
Let U_1 and U_2 be subspaces of the vector space V. Show that $U_1 \cap U_2$ is a subspace. Is $U_1 \cup U_2$ always a subspace ? If not, state conditions such that $U_1 \cup U_2$ is a subspace.   ☐

Exercise 27
Let V denote the set of all real $n \times n$-matrices.
Show that V with the usual scalar multiplication and addition is a vector space.
Is the set of all *regular* $n \times n$-matrices a subspace of V?
Is the set of all *symmetric* $n \times n$-matrices a subspace of V?
☐

Exercise 28
In the space $C([a; b])$ we consider the sets
$U_1 =$ the set of polynomials defined on $[a; b]$.
$U_2 =$ the set of polynomials defined on $[a; b]$ of degree $\le n$.
$U_3 =$ the set of polynomials defined on $[a; b]$ of degree $= n$.
$U_4 =$ the set of all $f \in C([a; b])$ with $f(a) = f(b) = 0$.
$U_5 = C^1([a; b])$.
Which of the U_i, $i = 1, 2, ..., 5$ are subspaces of $C([a; b])$?
☐

Exercise 29
In $C([-1; 1])$ we consider the sets U_1 and U_2 consisting of the odd and even functions in $C([-1; 1])$, respectively.
Show that U_1 and U_2 are subspaces and that $U_1 \cap U_2 = \{0\}$.
Show that every $f \in C([-1; 1])$ can be written in the form $f = f_1 + f_2$, where $f_1 \in U_1$ and $f_2 \in U_2$, and that this decomposition is unique.
☐

Exercise 30

In the space $C^1([a;b])$ we have the norm

$$\|f\|_\infty = \sup_{t\in[a;b]} |f(t)|.$$

Show that we could take $\sup_{t\in(a;b)} |f(t)|$, instead.
Show that $C^1([a;b])$ with the sup-norm is not a Banach space.
Show that

$$\|f\|_\infty^* = \sup_{t\in[a;b]} |f(t)| + \sup_{t\in[a;b]} |f'(t)|$$

is also a norm on $C^1([a;b])$ and that it is a Banach space with this norm.
☐

Exercise 31

Let $f \in C([a;b])$ and consider the p-norms

$$\|f\|_p = \left(\int_a^b |f(t)|^p dt\right)^{\frac{1}{p}}, \quad p \geq 1,$$

and

$$\|f\|_\infty = \sup_{t\in[a;b]} |f(t)|.$$

Show that $\|f\|_p \to \|f\|_\infty$ for $p \to \infty$.   ☐

Exercise 32

Show that a closed subspace of a Banach space is itself a Banach space.
☐

Exercise 33

Let $V_i, i = 1, 2, ..., n$ be normed vector spaces, with norms $\|\cdot\|_i, i = 1, 2, ..., n$.
The product space $V_1 \times V_2 \times ... \times V_n = \bigotimes_{i=1}^n V_i$ is defined by

$$\bigotimes_{i=1}^n V_i = \{(x_1, x_2, ..., x_n) \mid x_i \in V_i, i = 1, 2, ..., n\}.$$

In $\bigotimes_{i=1}^n V_i$ we use coordinatewise addition:

$$(x_1, x_2, ..., x_n) + (y_1, y_2, ..., y_n) = (x_1 + y_1, x_2 + y_2, ..., x_n + y_n),$$

and scalar multiplication:

$$\lambda(x_1, x_2, ..., x_n) = (\lambda x_1, \lambda x_2, ..., \lambda x_n),$$

and we define the norm by

$$\|(x_1, x_2, ..., x_n)\| = \sum_{i=1}^{n} \|x_i\|_i.$$

Show that $\bigotimes_{i=1}^{n} V_i$ with this norm is a normed vector space, and show that if all the spaces V_i with their repective norms are Banach spaces, then $\bigotimes_{i=1}^{n} V_i$ is a Banach space. ▯

Exercise 34
Assume that V and U are normed spaces and $f : V \rightarrow U$ is a continuous mapping, and assume that $X \subset V$ is a compact subset. Show that the image $f(X) \subset U$ is compact.
Show that a real function attains both maximum and minimum on a compact set. ▯

Exercise 35
Let V be a normed vector space and let $x_1, ..., x_k$ be k linearly independent vectors from V. Show that there exists a positive constant m such that for all scalars $\alpha_i \in C, i = 1, ..., k$ we have

$$\|\alpha_1 x_1 + ... + \alpha_k x_k\| \geq m(|\alpha_1| + ... + |\alpha_k|).$$

▯

Exercise 36
Show that any finite dimensional subspace of a normed vector space is a Banach space. ▯

Exercise 37
Let V be a vector space and let $\| \cdot \|$ and $\|| \cdot \||$ be two norms on V. The norms are said to be *equivalent* if there are positive constants m and M such that

$$m\|x\| \leq \||x\|| \leq M\|x\|$$

for all $x \in V$.
Show that all norms on a finite dimensional vector space are equivalent.
Show that equivalent norms define the same closed sets. ▯

Exercise 38
Show that a compact set in a normed vector space V is closed and bounded. If V is finite dimensional, show that a closed and bounded set is compact. ▯

Exercise 39

(Riesz' Lemma)

Let V be a normed vector space and let U be a closed subspace of V, $U \neq V$. Let α, $0 < \alpha < 1$ be given. Show that there is a $v \in V$ such that

$$\|v\| = 1 \quad \text{and} \quad \|v - u\| \geq \alpha$$

for all $u \in U$. ▯

Exercise 40

Let V be a Banach space. A series $\sum_{k=0}^{\infty} x_k$, $x_k \in V$ is *convergent* if the sequence (s_n), where

$$s_n = \sum_{k=0}^{n} x_k$$

is convergent in V.

Show that $\sum_{k=0}^{\infty} \|x_k\| < \infty$ implies that $\sum_{k=0}^{\infty} x_k$ is convergent.
Does the convergence of $\sum_{k=0}^{\infty} x_k$ imply that $\sum_{k=0}^{\infty} \|x_k\| < \infty$?
What if the space V is only assumed to be a normed space ?
▯

Exercise 41

In l^{∞}, the vector space of bounded sequences, we consider the sets U_1 and U_2, where U_1 denotes the set of sequences with only a finite number of elements different from 0, and U_2 the set of sequences with all but the N first elements different from 0.
Are U_1 and/or U_2 closed subspaces in l^{∞}?
Are U_1 and/or U_2 finite dimensional?   ▯

Exercise 42

Let T be a linear operator from a normed space V into a normed space W.
Show that the image $T(V)$ is a subspace of W.
Show that the kernel (or nullspace) $ker(T)$ is a subspace of of V.
If T is bounded, is it true that $T(V)$ and/or $ker(T)$ are closed?   ▯

Exercise 43

In the Banach space l^p, $1 \leq p \leq \infty$, we have a sequence (x_n) converging to an element x, where

$$x_n = (x_{n1}, x_{n2}, ...)$$

and

$$x = (x_1, x_2, ...).$$

Show that if $x_n \to x$ in l^p, then $x_{nk} \to x_k$ for all $k \in N$.
If $x_{nk} \to x_k$ for all $k \in N$, is it true that $x_n \to x$ in l^p? □

Exercise 44
Let T be a linear mapping from R^m to R^n, both equipped with the 2-norm.
Let (a_{ij}) denote a real $n \times m$ matrix corresponding to T. Show that T is
a bounded linear operator with $\|T\|^2 \le \sum_i \sum_j a_{ij}^2$. □

Exercise 45
Let T be a linear operator from a normed space V into a normed space W,
and assume that V is finite dimensional.
Show that T must be bounded. □

Exercise 46
Let T be a linear operator from a finite dimensional vector space into itself.
Show that T is injective if and only if T is surjective. □

Exercise 47
Let T be the linear mapping from $C^\infty(R)$ into itself given by $Tf = f'$.
Show that T is surjective.
Is T injective ? □

Exercise 48
Let $I = [a; b]$ be a bounded interval and consider the linear mapping T from
$C([a; b])$ into itself, given by

$$Tf(t) = \int_a^t f(s)ds.$$

We assume that $C([a; b])$ is equipped with the sup-norm.
Show that T is bounded and find $\|T\|$.
Show that T is injective and find $T^{-1} : T(C([a; b])) \to C([a; b])$.
Is T^{-1} bounded ? □

Exercise 49
Let T be a bounded linear operator from a normed vector space V into
a normed vector space W, and assume that T is surjective. Assume that
there is a $c > 0$ such that
$$\|Tx\| \ge c\|x\|$$
for all $x \in V$.
Show that T^{-1} exists and that $T^{-1} \in B(W, V)$. □

Exercise 50
Prove that in a *real* vector space with inner product we have

$$(x, y) = \frac{1}{4}(\|x + y\|^2 - \|x - y\|^2),$$

and in a *complex* vector space with inner product we have

$$(x, y) = \frac{1}{4}(\|x + y\|^2 - \|x - y\|^2 + i\|x + iy\|^2 - i\|x - iy\|^2).$$

These are the so-called *Polarization identities*. They tell us that, in a Hilbert space, the inner product is determined by the norm.   ☐

Exercise 51
Let V be a *real* normed vector space, and assume that the norm satisfies

$$\|x + y\|^2 + \|x - y\|^2 = 2(\|x\|^2 + \|y\|^2)$$

for all $x, y \in V$.
Show that

$$(x, y) = \frac{1}{4}(\|x + y\|^2 - \|x - y\|^2)$$

defines an inner product in V and that the norm is induced by this inner product.   ☐

Exercise 52
Show that the sup-norm on $C([a; b])$ is not induced by an inner product.
☐

Exercise 53
Prove that in a *real* vector space with inner product we have that $\|x\| = \|y\|$, implying that $(x + y, x - y) = 0$.
In the case $V = R^2$, this is a well-known geometric statement, which?   ☐

Exercise 54
Let V_i, $i = 1, ..., k$ be vector spaces equipped with inner products $(\cdot, \cdot)_i$, respectively. We define the product space $\bigotimes_{i=1}^{k} V_i$, as in Exercise 33. Show that we can define an inner product in $\bigotimes_{i=1}^{n} V_i$ by

$$((x_1, x_2, ..., x_k, y_1, y_2, ..., y_k)) = \sum_{i=1}^{k}(x_i, y_i)_i,$$

and that $\bigotimes_{i=1}^{n} V_i$ with this inner product is a Hilbert space if V_i, $i = 1, ..., k$ are Hilbert spaces.   ☐

Exercise 55
Let x and y be vectors in a vector space with an inner product. Show that $(x, y) = 0$ if and only if

$$\|x + \alpha y\| = \|x - \alpha y\|$$

for all scalars α.
Moreover, show that $(x, y) = 0$ if and only if

$$\|x + \alpha y\| \geq \|x\|$$

for all scalars α. ▯

Exercise 56
Let x and y be vectors in a *complex* vector space with an inner product, and assume that
$$\|x + y\|^2 = \|x\|^2 + \|y\|^2.$$
Does this imply that $(x, y) = 0$? ▯

Exercise 57
Let V be a vector space with an inner product and assume that $T \in B(V)$. Show that $(Tx, y) = 0$ for all $x, y \in V$ if and only if T is the zero operator. Show next that $(Tx, x) = 0$ for all $x, \in V$ if and only if T is the zero operator.
If the vector space is assumed to be real, do these results hold?
▯

Exercise 58
Let $[a; b]$ be a finite interval.
Show that $L^2([a; b]) \subset L^1([a; b])$. ▯

Exercise 59
Let T be a linear operator $T : L^2(R) \to L^2(R)$ satisfying that $f \geq 0$ implies that $Tf \geq 0$.
Show that
$$\|T(|f|)\| \geq \|Tf\|$$
for all $f \in L^2(R)$.
Show that T is bounded.
▯

Exercise 60
Let (e_n) be an orthonormal basis for the Hilbert space H.

Show that

$$T(\sum_{i=1}^{\infty} a_i e_i) = (a_1, a_2, ...)$$

defines an isomorphism from H onto l^2, satisfying $(Tx, Ty) = (x, y)$ for all $x, y \in H$. ∎

Exercise 61
Let M be a subset of a Hilbert space H. Show that $M^{\perp}$ is a closed subspace of H.
Show that $M \subset (M^{\perp})^{\perp}$, and show that $(M^{\perp})^{\perp}$ is the smallest closed subspace containing M.
∎

Exercise 62
Let (x_n) be an orthogonal sequence in a Hilbert space H, satisfying that

$$\sum_{n=1}^{\infty} ||x_n||^2 < \infty$$

Show that the series $\sum_{n=1}^{\infty} x_n$ is convergent in H.
Is this still true if we drop the orthogonality assumption? ∎

Exercise 63
Let H be a Hilbert space (infinite dimensional). Show that there is a sequence of vectors (x_n) such that $||x_n|| = 1$ for all n, and $(x_n, x) \to 0$ for all $x \in H$.
∎

Exercise 64
Let H be a Hilbert space. Show that

$$||x - z|| = ||x - y|| + ||y - z||$$

if and only if $y = \alpha x + (1 - \alpha)z$ for some $\alpha \in [0; 1]$.
∎

Exercise 65
In Chapter 4 we saw that any real, continuous function defined on $[0; \pi]$ can be approximated uniformly by linear combinations of cosines. Is the same true if we approximate with sines?
∎

Exercise 66
Let (e_n) be an orthonormal basis for $L^2([0;1])$. Construct from this an orthonormal basis for $L^2(I)$, where I is a finite interval. ⬚

Exercise 67
Let (e_n) be an orthonormal sequence in $L^2(I)$, where I is a finite interval with the property that for any *continuous* $f \in L^2(I)$ and any $\epsilon > 0$ we can find $N \in N$ and constants $a_1, a_2, ..., a_N$ such that

$$\|f - \sum_{k=1}^{N} a_k e_k\| < \epsilon.$$

Show that (e_n) is an orthonormal basis for $L^2(I)$.
⬚

Exercise 68
Let (e_n) be an orthonormal sequence in $L^2(I)$ where I is a finite interval with the property that
$$(e_n, f) = 0$$
for all n, and all *continuous* $f \in L^2(I)$ implies that $f = 0$.
Show that (e_n) is an orthonormal basis for $L^2(I)$. ⬚

Exercise 69
Prove that
$$\|P_n\|^2 = \frac{2}{2n+1}, \quad \text{for} \quad n = 0, 1, 2, ...$$
where P_n are the Legendre polynomials and the space is $L^2([-1;1])$.
⬚

Exercise 70
Show that the Legendre polynomials are orthogonal in $L^2([-1;1])$, and show that the set of *even* normalized Legendre functions (p_n), $n = 0, 2, 4, ...$ is an orthonormal basis for the closed subspace of even functions in $L^2([-1;1])$. By the way, why is this subspace closed?
⬚

Exercise 71
Show that the Hermite polynomium $H_n(t)$ can be written in the form

$$H_n(t) = n! \sum_{k=0}^{[\frac{n}{2}]} (-1)^k \frac{2^{n-2k}}{k!(n-2k)!} t^{n-2k},$$

where $[x]$ denotes the integer part of x.
Show that

$$H_{n+1}(t) = 2tH_n(t) - H'_n(t),$$

and that

$$H'_n(t) = 2nH_{n-1}(t).$$

Use these results to show that H_n is a solution to the *Hermite differential equation*:

$$\frac{d^2x}{dt^2} - 2t\frac{dx}{dt} + 2nx = 0, \quad t \in R.$$

☐

Exercise 72

Consider in $L^2([0;1])$ the sequence of *Rademacher-functions*:

$$e_n(t) = \sum_{j=0}^{2^n-1} (-1)^j 1_{]\frac{j}{2^n};\frac{j+1}{2^n}]}(t), \quad n \in N.$$

1: Draw the graphs for e_1, e_2, e_3, and e_4.
2: Show that (e_n) is an orthonormal sequence in $L^2([0;1])$.
3: Show that (e_n) is not an orthonormal basis.

☐

Exercise 73

Consider in $L^2([0;1])$ the sequence of *Haar-functions*:

$$h_1(t) = 1$$

and

$$h_{2^m+k}(t) = \sqrt{2^m} \quad \text{for} \quad \frac{k-1}{2^m} \leq t \leq \frac{2k-1}{2^{m+1}}$$

$$h_{2^m+k}(t) = -\sqrt{2^m} \quad \text{for} \quad \frac{2k-1}{2^{m+1}} \leq t \leq \frac{k}{2^m}$$

$$h_{2^m+k}(t) = 0 \quad \text{else},$$

where $k = 1, 2, ..., 2^m$ and $m = 0, 1, 2, ...$
1: Sketch the graphs of $h_1, h_2, ..., h_8$.
2: Show that (h_n) is an orthonormal sequence in $L^2([0;1])$.
3: Show that (h_n) is an orthonormal basis in $L^2([0;1])$.   ☐

Exercise 74

Let H be a Hilbert space and let P and Q denote tha orthogonal projections on the closed subspaces M and N, respectively. Show that if $M \perp N$, then $P + Q$ is the orthogonal projection on $M \oplus N$.   ⬚

Exercise 75

Let P and Q denote orthogonal projections in a Hilbert space, and assume that $PQ = QP$. Show that $P + Q - PQ$ is an orthogonal projection and find the image of $P + Q - PQ$.   ⬚

Exercise 76

Consider $C([a; b])$ with the sup-norm, (Here we take only real functions and consider it as a real vector space) and consider the functionals

$$\alpha(f) = \max_{t \in [a;b]} f(t)$$

$$\beta(f) = \min_{t \in [a;b]} f(t).$$

Are these functionals linear and/or bounded?   ⬚

Exercise 77

Let φ denote a linear functional on a vector space V, and assume that $ker(\varphi) \neq V$. Let $x_0 \in V \setminus ker(\varphi)$. Show that any vector $x \in V$ can be written in the form $x = a x_0 + y$, where $y \in ker(\varphi)$. Is this expansion unique?   ⬚

Exercise 78

Let φ and ψ denote linear functionals on a vector space V, and assume that $ker(\varphi) = ker(\psi)$.
Show that there is a constant $\alpha \in C$ such that $\alpha \varphi = \psi$.
⬚

Exercise 79

Let V denote a normed vector space with norm $\|\cdot\|$. Recall that V^* denotes the vector space of bounded, linear functionals on V. On V^* we define the *dual norm* $\|\varphi\|^*$, which is just the operator norm of φ as an element of $B(V, C)$.
Let $x \in V$. Show that

$$g_x(\varphi) = \varphi(x), \quad \varphi \in V^*$$

determines an element $g_x \in V^{**}$.

Show that the mapping $x \to g_x$ is a linear and injective mapping from V to V^{**}, and that $\|g_x\|^{**} = \|x\|$.

If $x \to g_x$ is also surjective, V is said to be *reflexive*.

Show that a Hilbert space is reflexive. ⬚

Exercise 80

We consider the space of sequences (l^p), where $p \geq 1$. Let $y \in l^q$ where $\frac{1}{p} + \frac{1}{q} = 1$. (If $p = 1$ then $y \in l^\infty$, the space of bounded sequences).

Show that

$$x \to \sum_{i=1}^{\infty} x_i \bar{y}_i$$

defines an element $y^* \in (l^p)^*$ with norm $\|y^*\|^* = \|y\|_q$. ⬚

Exercise 81

Let φ denote a bounded, linear functional on a Hilbert space H, and assume that the domain $D(\varphi)$ is a proper subspace of H. Show that there is exactly one extension φ_1 of φ to H with the property that $\|\varphi_1\| = \|\varphi\|$. ⬚

Exercise 82

Let H be a Hilbert space. A mapping $h : H \times H \to C$ is called *sesquilinear* if, for all $x, x_1, x_2 \in H$ and $\alpha \in C$, we have:

$$h(x_1 + x_2, x) = h(x_1, x) + h(x_2, x),$$
$$h(x, x_1 + x_2) = h(x, x_1) + h(x, x_2),$$
$$h(\alpha x_1, x_2) = \alpha h(x_1, x_2),$$
$$h(x_1, \alpha x_2) = \bar{\alpha} h(x_1, x_2).$$

We say that h is *bounded* if there is a constant $c \geq 0$ such that

$$|h(x_1, x_2)| \leq c \|x_1\| \|x_2\|$$

for all $x_1, x_2 \in H$. The norm $\|h\|$ is defined as the smallest possible c.

Show that there is a $S \in B(H)$ such that

$$h(x_1, x_2) = (Sx_1, x_2),$$

and that this representation is unique. Show also that $\|h\| = \|S\|$.

A sesquilinear form is called *Hermitian* if

$$h(x, y) = \bar{h}(y, x)$$

for all $x, y \in H$. If, moreover, $h(x, x) \geq 0$, the form is called *positive semidefinite*.

Show that in this case we have *Schwarz' inequality:*

$$|h(x,y)|^2 \le h(x,x)h(y,y)$$

for all $x, y \in H$.

☐

Exercise 83
Let V and W be Hilbert spaces and let $T \in B(V, W)$. Show that the image of a weakly convergent sequence in V is a weakly convergent sequence in W.   ☐

Exercise 84
In the Hilbert space l^2, we define an operator $T : D(T) \to l^2$ by

$$T((x_n)) = (a_n x_n),$$

where (a_n) is a complex sequence.
Find the maximal possible $D(T)$ and show that T is linear. Show that $D(T)$ is dense in l^2.
Show that if (a_n) is bounded, then $D(T) = l^2$ and T is bounded.

☐

Exercise 85
Consider in $L^2(R)$ the operator Q defined by

$$Qf(x) = xf(x),$$

with

$$D(Q) = \{f \in L^2(R) \mid Qf \in L^2(R)\}.$$

Show that Q is linear but not bounded. Show that $D(Q)$ is dense in $L^2(R)$. In quantum mechanics, Q is called the *position operator.*   ☐

Exercise 86
Consider in $L^2(R)$ the operator P defined by

$$Pf = -i\frac{\partial f}{\partial x},$$

with

$$D(P) = f \in L^2(R) \mid Pf \in L^2(R).$$

Show that P is linear but not bounded. Show that $D(P)$ is dense in $L^2(R)$. In quantum mechanics, P is called the *momentum operator.*

◻

Exercise 87

Let V be a normed vector space. Show that no pair of operators $S, T \in B(V)$ satisfies the *cannonical commutator relation*:

$$[S, T] = ST - TS = I.$$

(Hint: Show by induction that $ST^n - T^n S = nT^{n-1}$, $n \in N$, and use this to estimate $\|S\|$ and $\|T\|$.)   ◻

Exercise 88

Let V be a normed space and assume that $T \in B(V)$ is bijective. Show that if T^{-1} is bounded, then

$$\|T^{-1}\| \geq \|T\|.$$

◻

Exercise 89

Let (e_n) denote an orthonormal basis in a Hilbert space H, and define the operator T by

$$T(\sum_{k=1}^{\infty} x_k e_k) = \sum_{k=1}^{\infty} x_k e_{k+1}.$$

Show that $T \in B(H)$ and find $\|T\|$.
Show that T is injective and find T^{-1}.   ◻

Exercise 90

Let (e_k) be an orthonormal basis in a Hilbert space H, and let $T \in B(H)$. Define for $j, k \in N$ the numbers

$$t_{jk} = (Te_j, e_k).$$

Show that

$$Te_j = \sum_{k=1}^{\infty} t_{jk} e_k,$$

and that $\sum_{k=1}^{\infty} |t_{jk}|^2 < \infty$ for $j \in N$.
The matrix (t_{jk}) is called the *matrix form* for T with respect to the orthonormal basis (e_k).
Let $A, B \in B(H)$ have the forms (a_{jk}) and (b_{jk}), respectively. Find the forms for $A + B$ and AB.   ◻

Exercise 91

Let (e_k) be an orthonormal basis in a Hilbert space H, and let $T : D(T) \to K$ be a linear operator from the Hilbert space H into the Hilbert space K. Show that if $e_k \in D(T)$ for all $k \in N$, then $D(T)$ is dense in H. ⬚

Exercise 92

Consider in l^2 the operator T defined as

$$T(x_n) = (nx_n).$$

Show that T is a closed, densely defined operator, and show that $T(D(T)) = l^2$. ⬚

Exercise 93

Let T be a closed linear operator $T : D(T) \subset V \to V$, where V is a Banach space, and let $A \in B(V)$.
Show that $A + T$ and TA are closed, linear operators. ⬚

Exercise 94

Consider in l^2 the operator T defined as

$$T(x_1, x_2, ...) = (x_2, 2x_3, 3x_4, ...).$$

Show that T is a closed, densely defined operator. ⬚

Exercise 95

Let X, Y be Banach spaces, and let $T : D(T) \subset X \to Y$ be a closed linear operator. Assume that the sequences $(u_n), (v_n) \subset D(T)$ satisfy that $\lim_n u_n = \lim_n v_n$.
Show that if (Tu_n) and (Tv_n) are both convergent, then $\lim_n Tu_n = \lim_n Tv_n$. ⬚

Exercise 96

Let $T : X \to Y$ be a closed linear operator between two normed spaces, and let $A \subset X$ be compact. Show that $T(A)$ is closed. ⬚

Exercise 97

Let $T : D(T) \subset X \to Y$ be a closed linear operator between two normed spaces. Show that $ker(T)$ is a closed subspace of X. ⬚

Exercise 98

In the Hilbert space l^2 we consider the operator T given by

$$T(x_1, x_2, ...) = (x_1, 2x_2, 3x_3, ...).$$

Show that T is a closed linear operator, and show that $T(D(T)) = l^2$. □

Exercise 99

Let T be a closed linear operator on a Hilbert space H, and let $A \in B(H)$. Show that $A + T$ and TA are closed linear operators. □

Exercise 100

In the Hilbert space l^2 we consider the operator T given by

$$T(x_1, x_2, ...) = (x_2, 2x_3, 3x_4, ...).$$

Show that T is a closed linear operator. □

Exercise 101

Assume that $T : D(T) \subset H \to K$ is a closed linear operator from the Hilbert space H to the Hilbert space K. Let (u_n) and (v_n) be sequences in $D(T)$, and assume that $\lim_n u_n = \lim_n v_n$.
Show that if (Tu_n) and (Tv_n) are both convergent, then $\lim_n Tu_n = \lim_n Tv_n$.
□

Exercise 102

Let X and Y denote normed spaces, and let $T : X \to Y$ be a closed linear operator.
Show that $T(A)$ is closed in Y if $A \subset X$ is compact. □

Exercise 103

Let X and Y denote normed spaces, and let $T : X \to Y$ be a closed linear operator. Show that $ker(T)$ is a closed subspace of X. □

Exercise 104

Define, for $h \in R$, the operator τ_h on $L^2(R)$ by

$$\tau_h f(x) = f(x - h).$$

Show that τ_h is bounded. □

Exercise 105
Consider in $L^2(R)$ the operator Q defined by

$$Qf(x) = xf(x),$$

with

$$D(Q) = \{f \in L^2(R) \mid Qf \in L^2(R)\}.$$

Determine $\rho(Q)$ and $\sigma_p(Q)$. ☐

Exercise 106
Let (e_n) denote an orthonormal basis in a Hilbert space H, and consider the operator

$$T(\sum_{k=1}^{\infty} a_k e_k) = \sum_{k=1}^{\infty} a_k e_{k+1}.$$

Determine $\|T\|$ and $\sigma(T)$. ☐

Exercise 107
In l^2 we consider the operator

$$(x_1, x_2, x_3, ...) \rightarrow (x_1, \frac{1}{2}(x_1+x_2), \frac{1}{3}(x_1+x_2+x_3), ..., \frac{1}{2^{n-1}}(x_1+x_2+...+x_n), ...).$$

Show that the operator is bounded and not surjective. Let (e_n) denote an orthonormal basis in a Hilbert space H, and consider the operator

$$T(\sum_{k=1}^{\infty} a_k e_k) = \sum_{k=2}^{\infty} \sqrt{k} a_k e_{k-1}.$$

Determine the spectrum $\sigma(T)$, and find for each eigenvalue the corresponding eigenvectors. ☐

Exercise 108
Let (e_n) denote an orthonormal basis in a Hilbert space H. We define the sequence $(f_k)_{k \in Z}$ by

$$f_0 = e$$
$$f_k = e_{2k+1} \quad \text{for} \quad k > 0,$$
$$f_k = e_{-2k} \quad \text{for} \quad k < 0.$$

In this way $(f_k)_{k \in Z}$ is an orthonormal basis. We define the *doublesided shift operator* S by

$$S(\sum_{k=-\infty}^{\infty} a_k f_k) = \sum_{k=-\infty}^{\infty} a_k f_{k+1}.$$

Show that S is a bounded operator, and show that S has no eigenvalues.
□

Exercise 109
Define, for $h \in R$, the operator τ_h on $L^2(R)$ by

$$\tau_h f(x) = f(x - h).$$

Show that τ_h has no eigenvalues and that

$$\sigma(\tau_h) \subset \{z \in C \mid |z| = 1\}.$$

(It is, in fact, true that $\sigma(\tau_h) = \{z \in C \mid |z| = 1\}$).   □

Exercise 110
Let $T \in B(H)$ where H is a Hilbert (or just Banach) space. Show that $\|R_\lambda(T)\| \to 0$ for $|\lambda| \to \infty$.   □

Exercise 111
Let T be a self-adjoint operator in a Hilbert space H. Show that if $D(T) = H$, then T is bounded.   □

Exercise 112
Let T be a bounded operator on a Hilbert space H, and assume that N and M are closed subspaces of H.
Show that

$$T(M) \subset N$$

if and only if

$$T^*(N^\perp) \subset M^\perp.$$

Show, moreover, that

$$ker(T) = T^*(H)^\perp$$

and

$$ker(T)^\perp = \overline{T^*(H)}.$$

□

Exercise 113
Let T be a bounded operator on a Hilbert space H with $\|T\| = 1$, and assume that we can find $x_0 \in H$ such that $Tx_0 = x_0$. Show that also $T^*x_0 = x_0$.   □

Exercise 114

Let (e_n) denote an orthonormal basis in a Hilbert space H, and consider the operator

$$T(\sum_{k=1}^{\infty} a_k e_k) = \sum_{k=1}^{\infty} a_k e_{k+1}.$$

Find the adjoint T^* and show that T^* is an extension of T^{-1}.

▯

Exercise 115

Let (e_n) denote an orthonormal basis in a Hilbert space H, and consider the operator

$$T(\sum_{k=1}^{\infty} a_k e_k) = \sum_{k=2}^{\infty} \sqrt{k-1} a_k e_{k-1}.$$

Show that T is a densely defined, unbounded operator, and find T^*.   ▯

Exercise 116

Let $T \in B(H)$. Show that we can write T as

$$T = A + iB$$

where A and B are uniquely determined, bounded, self-adjoint operators.

▯

Exercise 117

Show that $T \in B(H)$ is self-adjoint if and only if one of the following conditions are satisfied:

$$(Tx, x) = (x, Tx) \quad \text{for all} \quad x \in H$$

or

$$(Tx, x) \in R \quad \text{for all} \quad x \in H.$$

▯

Exercise 118

Let S and T be bounded, self-adjoint operators on a Hilbert space. Show that $ST + TS$ and $i(ST - TS)$ are self-adjoint.   ▯

Exercise 119

Let T be a bounded self-adjoint operator. Define the numbers

$$m = \inf\{(Tx, x) \mid \|x\| = 1\}$$

and
$$M = \sup\{(Tx, x) \mid \|x\| = 1\}.$$

Show that $\sigma(T) \subset [m; M]$, and show that both m and M belong to $\sigma(T)$. Show that $\|T\| = \max\{|m|, |M|\}$.   ⬚

Exercise 120
Consider in $L^2(R)$ the operator Q defined by

$$Qf(x) = xf(x),$$

with

$$D(Q) = \{f \in L^2(R) \mid Qf \in L^2(R)\}.$$

Show that Q is self-adjoint.   ⬚

Exercise 121
Show that the set of self-adjoint operators is closed in $B(H)$.   ⬚

Exercise 122
Let (e_n) denote an orthonormal basis in a Hilbert space H, and let (r_k) be all the rational numbers in $]0; 1[$ arranged as a sequence. Consider the operator

$$T(\sum_{k=1}^{\infty} a_k e_k) = \sum_{k=1}^{\infty} r_k a_k e_k.$$

Show that T is self-adjoint and that $\|T\| = 1$. Find $\rho(T)$ and determine the point spectrum and the continuous spectrum for T.   ⬚

Exercise 123
Let $T \in B(H)$. An operator is *isometric* if $\|Tx\| = \|x\|$ for all $x \in H$. Show that the following conditions are equivalent for $T \in B(H)$.
a: T is isometric.
b: $T^*T = I$.
c: $(Tx, Ty) = (x, y)$ for all $x, y \in H$.   ⬚

Exercise 124
Let $T \in B(H)$ be an isometric operator. Show that $T(H)$ is a closed subspace. Show that $T(H) = H$ if H is finite dimensional. Give an example of an isometric operator with $T(H) \neq H$.   ⬚

Exercise 125

Let $T \in B(H)$ be an isometric operator, and let M and N denote closed subspaces of the Hilbert space H. Show that

$$T(M) = N \Rightarrow T(M^\perp) \subset N^\perp.$$

Show that T is isometric if and only if, for any orthonormal basis, (e_k), (Te_k) is an orthonormal sequence. ☐

Exercise 126

Let $T \in B(H)$ be an isometric operator. Show that TT^* is a projection and determine the range. ☐

Exercise 127

Consider the Hilbert space $L^2([0; \infty))$. Let $h > 0$ and define the operator T by

$$Tf(x) = 0 \quad \text{for} \quad 0 \le x < h,$$
$$Tf(x) = f(x - h) \text{for} \quad h \le x.$$

Show that T is isometric and determine T^*. Find TT^* and T^*T. ☐

Exercise 128

An operator $T \in B(H)$ is called *unitary* if it is isometric and surjective. Show that the following conditions are equivalent for an operator $T \in B(H)$.
a: T is unitary.
b: T is bijective and $T^{-1} = T^*$.
c: $T^*T = TT^* = I$.
d: T and T^* are isometric.
e: T is isometric and T^* is injective.
f: T^* is unitary. ☐

Exercise 129

Let (e_k) denote an orthonormal basis in a Hilbert space H, and let $T \in B(H)$ be given by

$$T(\sum_{k=1}^{\infty} a_k e_k) = \sum_{k=1}^{\infty} \lambda_k a_k e_k.$$

Show that T is unitary if and only if $|\lambda_k| = 1$ for all k. ☐

Exercise 130

Let $T \in B(H)$ be unitary. Show that

$$\sigma(T) \subset \{z \in C \mid |z| = 1\}.$$

⬚

Exercise 131

An operator $T \in B(H)$ is *normal* if

$$TT^* = T^*T.$$

Show that T is normal if and only if $\|T^*x\| = \|Tx\|$ for all $x \in H$

⬚

Exercise 132

Let $T \in B(H)$ be normal. Show that

$$\|(T - \lambda I)x\| = \|(T^* - \bar{\lambda}I)x\|$$

for all $x \in H$. Show that $\sigma_r(T)$ is empty.   ⬚

Exercise 133

An operator $T \in B(H)$ is *positive* if

$$(Tx, x) \geq 0$$

for all $x \in H$, and we write $T \geq 0$.
Prove the following:
a: $T \geq 0$ implies that T is self-adjoint.
b: If $S, T \geq 0$, $\alpha \geq 0$, then $S + \alpha T \geq 0$.
c: If $T \geq 0$ and $S \in B(H)$, then $S^*TS \geq 0$.
d: If $T \in B(H)$, then $T^*T \geq 0$.
e: If T is an orthogonal projection, then $T \geq 0$.   ⬚

Exercise 134

Let P_M and P_N denote the orthogonal projections on the closed subspaces M and N of a Hilbert space H. Show that $M \subset N$ implies that $P_M \leq P_N$.
⬚

Exercise 135

An operator $T \in B(H)$ is called a *contraction* if

$$\|Tx\| \leq \|x\|$$

for all $x \in H$.
Show that the following conditions are equivalent for an operator $T \in B(H)$:
a: T is a contraction

b: $\|T\| \leq 1$
c: $T^*T \leq I$
d: $TT^* \leq I$
e: T^* is a contraction
f: T^*T is a contraction $\quad\Box$

Exercise 136
Let S and T be linear and bounded operators, and assume that S is compact. Show that ST and TS are compact.
$\Box$

Exercise 137
Let S and T be compact operators in $B(H)$, and let $\alpha \in C$. Show that $S + \alpha T$ is compact. $\quad\Box$

Exercise 138
Let (e_k) denote an orthonormal basis in a Hilbert space H, and define the operator T by

$$T(\sum_{k=1}^{\infty} a_k e_k) = \sum_{k=2}^{\infty} \frac{1}{k} a_k e_{k-1}.$$

Show that T is compact and find T^*. Find $\sigma_p(T)$ and $\sigma_p(T^*)$. $\quad\Box$

Exercise 139
Let T be a bounded operator on a Hilbert space H. Show that:
a: If T is compact, then T^* is also compact.
b: If T^*T is compact, then T is compact.
c: If T is self-adjoint and T^n is compact for some n, then T is compact.
$\Box$

Exercise 140
Let (e_k) denote an orthonormal basis in a Hilbert space H, and assume that the operator T has the matrix representation (t_{jk}) with respect to the basis (e_k). Show that

$$\sum_{j=1}^{\infty} \sum_{k=1}^{\infty} |t_{jk}|^2 < \infty$$

implies that T is compact.
Let (f_k) denote another orthonormal basis in H, and let $s_{jk} = (Tf_j, f_k)$ so that (s_{jk}) is the matrix representation of T with respect to the basis (f_k).

Show that

$$\sum_{j=1}^{\infty}\sum_{k=1}^{\infty}|t_{jk}|^2 = \sum_{j=1}^{\infty}\sum_{k=1}^{\infty}|s_{jk}|^2.$$

An operator satisfying $\sum_{j=1}^{\infty}\sum_{k=1}^{\infty}|t_{jk}|^2 < \infty$ is called a *general Hilbert-Schmidt operator.* ◻

Exercise 141
For a general Hilbert-Schmidt operator, we define the Hilbert-Schmidt norm $\|\cdot\|_{HS}$ by

$$\|T\|_{HS} = (\sum_{j=1}^{\infty}\sum_{k=1}^{\infty}|t_{jk}|^2)^{\frac{1}{2}}.$$

Show that this *is* a norm, and show that

$$\|T\| \le \|T\|_{HS}$$

for a general Hilbert-Schmidt operator T. ◻

Exercise 142
Define for $f \in L^2(R)$ the operator K by

$$Kf(x) = \int_{-\infty}^{\infty} \frac{1}{2}e^{-|x-t|}f(t)dt.$$

Show that $Kf \in L^2(R)$, and that K is linear and bounded with norm ≤ 1. Show that the function $\frac{1}{2}e^{-|x-t|}$ does not belong to $L^2(R^2)$, so that K is not a Hilbert-Schmidt operator. ◻

Exercise 143
Let K denote the Hilbert-Schmidt operator with kernel

$$k(x,t) = \sin(x)\cos(t), \quad 0 \le x,t \le 2\pi.$$

Show that the only eigenvalue for K is 0.
Find an orthonormal basis for $ker(K)$.
◻

Exercise 144
Let K denote the Hilbert-Schmidt operator with continuous kernel k on $L^2(I)$ where I is a closed and bounded interval. Show that all the iterated kernels k_n are continuous on I^2, and show that

$$\|k_n\|_2 \le \|k\|_2^n.$$

Show that if $|\lambda|\|k\|_2 < 1$, then the series

$$\sum_{n=1}^{\infty} \lambda^n k_n$$

is convergent in $L^2(I)$. ☐

Exercise 145
Let K and L denote the Hilbert-Schmidt operators with continuous kernels k and l on $L^2(I)$ where I is a closed and bounded interval. We define the *trace* of K, $tr(K)$ by

$$tr(K) = \int_I k(x,x)dx,$$

and similarly for L.
Show that

$$|tr(KL)| \leq \|K\|_{HS}\|L\|_{HS}$$

and

$$|tr(K^n)| \leq \|K\|_{HS}^n, \quad n \geq 2.$$

Moreover, if $(K_n), (L_n)$ denote sequences of Hilbert-Schmidt operators like above, where

$$\|K_n - K\|_{HS} \to 0 \quad \text{and} \quad \|L_n - L\|_{HS} \to 0,$$

then

$$tr(K_n L_n) \to tr(KL).$$

☐

Exercise 146
Let K denote the Hilbert-Schmidt operator on $L^2([0;1])$ with kernel

$$k(x,t) = x + t.$$

Find all eigenvalues and eigenfunctions for K.
Solve the equation $Ku = \mu u + f$, $f \in L^2([0;1])$ when μ is not in the spectrum for K.
☐

Exercise 147
Let K denote the Hilbert-Schmidt operator on $L^2([-\frac{\pi}{2}; \frac{\pi}{2}])$ with kernel

$$k(x,t) = \cos(x - t).$$

Find all eigenvalues and eigenfunctions for K.

Solve the equation $Ku = \mu u + f$, $f \in L^2([-\frac{\pi}{2}; \frac{\pi}{2}])$ when μ is not in the spectrum for K. ☐

Exercise 148

Let K denote the Hilbert-Schmidt operator on $L^2([-\pi; \pi])$ with kernel

$$k(x, t) = (\cos(x) + \cos(t))^2.$$

Find all eigenvalues and eigenfunctions for K, and find an orthonormal basis for $ker(K)$. ☐

Exercise 149

Let K denote a self-adjoint Hilbert-Schmidt operator on $L^2(I)$ with kernel k.

Show that $\|K\| = \|k\|_2$ if and only if the spectrum for K consists of only two points. ☐

━━━

Examination Sets

The following exercises are four-hour examination sets posed in the course "Funktional Analyse" in the period 1979-1989.

Some of the exercises have been edited slightly in order to accomplish the notation and background given by the present book.

Winter 79/80

Exercise 150

Let $T : l^2 \to l^2$ denote the operator

$$T(x_1, x_2, ..., x_n, ...) = (x_2, x_4, ..., x_{2n}, ...).$$

Find $\|T\|$.

Find all eigenvalues for T.

Show that the eigenspace corresponding to any eigenvalue is infinite-dimensional.

Determine the operators T^*, TT^*, and T^*T.

Determine $\sigma(T)$ and $\rho(T)$. ☐

Exercise 151

We will consider $H = L^2([0;1])$ as a *real* Hilbert space, and define $T : H \to H$ by

$$Tf(x) = \int_0^x f(t)dt.$$

Show that

$$|Tf(x)| \le \sqrt{x}\|f\|_2,$$

and use this to show that $\|T\| < 1$.
Show that

$$T^n f(x) = \int_0^x \frac{(x-t)^{n-1}}{(n-1)!} f(t)dt.$$

Show that $\log(I + T)$ is a well-defined operator of Volterra type, and find an explicit expression for the kernel of this operator using only known functions, that is, find k such that

$$\log(I + T)f(x) = \int_0^x k(x,t)f(t)dt.$$

□

Exercise 152

Let $k(x,t) = (\sin(x) + \sin(t))^2 - \frac{1}{8}$ be the kernel for a Hilbert-Schmidt operator K on the complex Hilbert space $L^2([-\pi;\pi])$.
Show that K is self-adjoint and express the range $K(L^2([-\pi;\pi]))$ of K with the help of the non-normalized basis

$$1, \cos(x), \sin(x), \cos(2x), \sin(2x), \dots.$$

Find all nonzero eigenvalues and corresponding eigenfunctions for K, and determine $\sigma(K)$.
Solve the equation $Ku = \pi u - \frac{5\pi}{4}$ in $L^2([-\pi;\pi])$.   □

Winter 81/82

Exercise 153

Let V be a Banach space and let $T \in B(V)$ such that T^{-1} exists and belongs to $B(V)$.
Show that if $\|T\| \le 1$ and $\|T^{-1}\| \le 1$, then

$$\|T\| = \|T^{-1}\| = 1,$$

and $\|Tf\| = \|f\|$ for all $f \in V$.   □

Exercise 154
Let H be a Hilbert space and let $T \in B(H)$ be positive and self-adjoint. Show that
$$|(Tx, y)|^2 \leq (Tx, x)(Ty, y)$$
for all $x, y \in H$. ▯

Exercise 155
Consider the Hilbert space $H = L^2([-\pi; \pi])$ with the orthonormal basis (e_n), defined by
$$e_n(t) = \frac{1}{\sqrt{2\pi}} e^{int}, \quad n \in Z.$$

For $f \in H$ we define $\widehat{f}$ by
$$\widehat{f}(x) = \frac{1}{2\pi} \int_{-\pi}^{\pi} f(t) e^{-ixt} dt.$$

Show that $\widehat{f}$ exists for all $x \in R$.
Use the function $\widehat{f}$ to express the Fourier expansion of $f \in H$ in terms of the orthogonal basis $(\sqrt{2\pi} e_n)$.
Let $\gamma \in R$ and $f \in H$ be given and define the function g by
$$g(t) = f(t) e^{-i\gamma t}.$$

Find $\widehat{g}(x)$.
Show that for any $\gamma \in R$ and $f \in H$ we have
$$\sum_{n=-\infty}^{\infty} |\widehat{f}(n + \gamma)|^2 = \frac{1}{2\pi} \|f\|_2^2.$$

Take $f = 1$ and $\gamma = \frac{\theta}{\pi}$, $\theta \notin \{p\pi \mid p \in Z\}$ and show that
$$\frac{1}{\sin^2(\theta)} = \sum_{n=-\infty}^{\infty} \frac{1}{(n\pi + \theta)^2}.$$

▯

Exercise 156
Let $k(x, t) = x + t + 2xt$ be the kernel for the Hilbert-Schmidt operator K on the Hilbert space $H = L^2([-1; 1])$.
Show that K is self-adjoint and determine the range $K(H)$.
Find all nonzero eigenvalues and corresponding eigenfunctions for K, and determine $\sigma(K)$ as well as $\|K\|$.

Express Kf, $f \in H$ with the help of the Legendre polynomials (P_n).
Let $f(x) = \cosh(1)\cosh(x) - \cosh(2x)$. Show that $(f, P_0) = (f, P_1) = 0$,
and solve the equation

$$Ku(x) + u(x) = f(x).$$

◻

Winter 82/83

Exercise 157
This exercise consists of five small, independent exercises.

a) Let V denote a normed space. Show that $\|x - y\| \geq \|\|x\| - \|y\|\|$ for all
$x, y \in V$.
b) Let T be a bounded, linear, and self-adjoint operator on a Hilbert space.
Assume that T is surjective and show that T is then injective.
c) Assume that T is a closed linear operator on a normed space X. Show
that $ker(T)$ is closed in X.
d) Let H denote a Hilbert space and assume that (x_n) and (y_n) are two
sequences in the closed unit ball af H such that $(x_n, y_n) \to 1$. Show that
$\|x_n - y_n\| \to 0$.
e) Let (x_n) and (y_n) denote two orthonormal sequences in a Hilbert space
H, and assume that

$$\sum_{n=1}^{\infty} \|x_n - y_n\|^2 < 1.$$

Show that if (x_n) is an orthonormal basis, then so is (y_n) .   ◻

Exercise 158
Let H denote a Hilbert space and let $T \in B(H)$, and assume that there is
a positive c such that

$$|(Tx, x)| \geq c\|x\|^2$$

for all $x \in H$.
Show that T^{-1} exists and belongs to $B(H)$.   ◻

Exercise 159
Let H denote a Hilbert space and define for $a, b \in H$ the mapping $T_{a,b}$:
$H \to H$ by

$$T_{a,b}x = (x, b)a.$$

a) Show that $T_{a,b} \in B(H)$, and find $dim(T_{a,b}(H))$ (when $a, b \neq 0$) as well as the norm of $T_{a,b}$.
b) Find $T_{a,b}^*$.
c) Let $T \in B(H)$ have one-dimensional range. Show that there exists $a, b \in H$ such that $T = T_{a,b}$.   ▯

Exercise 160
In $L^2([-\pi; \pi])$ we consider the orthonormal basis (e_n), $n \in Z$ where

$$e_n(t) = \frac{1}{\sqrt{2\pi}} e^{int}.$$

a) Let $\varphi : R \to C$ denote a continuous function with period 2π, and assume that $\varphi(-x) = \overline{\varphi}(x)$ for all $x \in R$. Show that

$$Ku(x) = \int_{-\pi}^{\pi} \varphi(x - t)u(t)dt$$

defines a self-adjoint Hilbert-Schmidt operator on $L^2([-\pi; \pi])$.
b) Show that all e_n are eigenfunctions for K.
From now on we assume that φ is the periodic extension from $[-\pi; \pi]$ to R of the function

$$\varphi(x) = 1 - \frac{|x|}{\pi}.$$

c) Calculate the spectrum of K.
d) Solve the equation $Ku = \frac{2}{\pi}u + f$ in $L^2([-\pi; \pi])$, where $f(x) = \sin^2(x) + \sin(x)$.
e) Solve the equation $Ku = \frac{4}{\pi}u + 1$ in $L^2([-\pi; \pi])$.   ▯

Winter 83/84

Exercise 161
Let $(x_n) \in l^2$ and define the sequence $y = (y_n)$ by

$$y_n = x_{n+1} + nx_n + x_{n-1},$$

where we put $x_0 = 0$ whenever it is necessary.
1) Show that $y \in l^2$ if and only if $(nx_n) \in l^2$.
Let $D = \{x \in l^2 \mid (nx_n) \in l^2\}$ and define a linear operator $T : D \to l^2$, $Tx = y$, where y is given above.
2) Show that D is dense in l^2.
3) Show that T is self-adjoint.
   ▯

Exercise 162

Let X denote the Banach space of $C([-1;1])$-functions equipped with the usual sup-norm $\|\cdot\|_\infty$, and let $T \in B(X)$ be given by

$$Tf = f(0) + f.$$

1) Find the norm of T.
2) Determine the resolvent set $\rho(T)$ for T, and find $T_\lambda^{-1} = (T - \lambda I)^{-1}$ for all $\lambda \in \rho(T)$.
3) Show that the spectrum for T is a pure point spectrum, and find all eigenvalues and corresponding eigenfunctions.
4) Show that all $f \in X$ can be written as a sum of eigenfunctions belonging to different eigenspaces, and show that this decomposition is unique. ⬚

Exercise 163

In this exercise it is allowed to change the order of integrations without justification.
Consider the operator

$$Af(x) = \frac{1}{\sqrt{\pi}} \int_0^x \frac{f(t)}{\sqrt{x-t}} dt, \quad x \in [0;1]$$

whenever this expression gives sense.
1) Show that $Af \in L^\infty([0;1])$ if $f \in L^p([0;1])$, $p > 2$.
2) Find the operator $B = A^2$, that is, find the kernel $k(x,t)$ such that

$$Bf(x) = A^2 f(x) = \int_0^x k(x,t)f(t)dt$$

for $f \in L^p([0;1])$, $p > 2$.
3) Show that $B : L^p([0;1]) \to L^\infty([0;1])$, $1 \le p \le \infty$ is bounded.
4) Solve the equation

$$(I - A)f(x) = 1$$

formally by a Neumann series, and express f as

$$f(x) = g(x) + Ah(x)$$

where g and h are known functions. (Here it is not possible to express $Ah(x)$ as known functions.)
Insert and show that this formal solution *is* a solution.
⬚

Exercise 164

Let $H = L^2([0;1])$, and consider the operator

$$Tf(x) = \sqrt{3}xf(x^3).$$

1) Show that $T \in B(H)$ and find $\|T\|$.
2) Show that T^{-1} exists and that $T^{-1} \in B(H)$: Determine $T^{-1}g(y)$ for $g \in H$, and find $\|T^{-1}\|$.
3) Show that $\sigma(T) \subset \{\lambda \in C \mid |\lambda| = \|T\|\}$. ▯

Winter 84/85

Exercise 165
Consider the operator $T : l^2 \to l^2$ given by

$$T(x_1, x_2, ..., x_n, ...) = (\frac{1}{2}x_2, \frac{2}{3}x_3, ..., \frac{n}{n+1}x_n, ...).$$

1) Determine $\|T\|$.
2) Find all eigenvalues $\sigma_p(T)$ and corresponding eigenvectors.
3) Determine the adjoint T^* and $\sigma_p(T^*)$ and the resolvent $\rho(T)$. ▯

Exercise 166
Let $w(t) \geq 0$ be a non-negative function on R. We define a linear functional I_w by

$$I_w(f) = \int_R f(t)w(t)dt,$$

for $fw \in L^1(R)$.
Assume that $|f|^p w$ and $|g|^q w$ are in $L^1(R)$, where f and g are (measurable) functions and $1 < p, q < \infty$ with $\frac{1}{p} + \frac{1}{q} = 1$.
1) Show the generalized Hölder's inequality

$$|I_w(fg)| \leq (I_w(|f|^p))^{\frac{1}{p}}(I_w(|g|^q))^{\frac{1}{q}},$$

where the inequality for $w = 1$ can be taken to be valid.
Now recall the Gamma function

$$\Gamma(x) = \int_0^\infty t^{x-1}e^{-t}dt, \quad x > 0,$$

with the property $\Gamma(x + 1) = x\Gamma(x)$ for $x > 0$.
Use the generalized Hölder's inequality with $w(t) = t^{n-1}e^{-t}1_{]0;\infty[}$ and $p = q = 1$ to show that

$$\Gamma(n + \frac{1}{2}) \leq \frac{n!}{\sqrt{n}}, \quad n \in N.$$

Give a similar estimation of $\Gamma(n + 1)$ by taking $w(t) = t^{n-\frac{1}{2}}e^{-t}1_{]0;\infty[}$ and $p = q = 2$, and deduce that

$$\frac{n!}{\sqrt{n + \frac{1}{2}}} \leq \Gamma(n + \frac{1}{2}) \leq \frac{n!}{\sqrt{n}}, \quad n \in N.$$

⧠

Exercise 167

Let $H = L^2([0; 1])$ and consider the integral operator

$$Bf(x) = \int_0^x f(t)dt$$

for $f \in H$.

a) Show that $k(x, t) = \min\{x, t\}$, $0 \le x, t \le 1$ is the kernel for the self-adjoint Hilbert-Schmidt operator $K = BB^*$.

b) Let φ be an eigenfunction for K associated with a nonzero eigenvalue λ. Justify that φ can be taken as a C^∞-function.

Next, show that φ must satisfy the equation $\lambda \varphi''(x) = -\varphi(x)$, and use this to find all nonzero eigenvalues for K and all the associated eigenfunctions.

c) Assuming that $\|BB^*\| = \|B^*\|^2$, show that $\|K\| = \|B\|^2$, and find both $\|K\|$ and $\|B\|$. ⧠

Winter 85/86

Exercise 168

Let $H = L^2([a, b])$, a, b finite, and T the operator

$$Tf(x) = \frac{1}{b-a} \int_a^b f(t)dt$$

for $x \in [a, b]$ and $f \in H$.

1) Show that $T \in B(H)$.

2) Show that T is a projection. ⧠

Exercise 169

Let $T : l^2 \to l^2$ be the linear operator given by

$$T(x_1, x_2, ..., x_n, ...) = (x_1 + x_2, x_2 + x_3, ..., x_n + x_{n+1}, ...).$$

1) Find the point spectrum $\sigma_p(T)$ and determine all eigenvectors associated to $\lambda \in \sigma_p(T)$.

2) Determine $\|T\|$.

3) Determine the adjoint T^*, and find also the point spectrum $\sigma_p(T^*)$.

4) Let $S = T - I$. Determine $\|S\|$.

5) Find $\sigma_c(T)$ and $\sigma_r(T)$ with the help of S above. ⧠

Exercise 170
Let H denote a Hilbert space and let $T \in B(H)$. Assume that for some $m \in N$ we have that $T^m = 0$.
Show that

$$(I - \lambda T)^{-1} = \sum_{n=0}^{m-1} \lambda^n T^n \in B(H),$$

and deduce that $C \setminus \{0\} \subset \rho(T)$. Show next that $\sigma(T) = \sigma_p(T) = \{0\}$.

□

Exercise 171
Let $H = L^2([-\pi; \pi])$ be the Hilbert space with the usual orthonormal basis

$$e_n(x) = \frac{1}{\sqrt{2\pi}} e^{inx}, \quad n \in Z.$$

Let $\lambda_n, n \in Z$ be complex numbers satisfying

$$M = \sup_{n \in Z} |\lambda_n - n| < \frac{\log 2}{\pi},$$

and let

$$f_n(x) = \frac{1}{\sqrt{2\pi}} e^{i\lambda_n x}, \quad n \in Z.$$

a) Show that

$$f_n(x) - e_n(x) = e_n(x) \sum_{k=1}^{\infty} \frac{(i(\lambda_n - n))^k}{k!} x^k, \quad n \in Z.$$

b) Show that for any $k \in Z$ we have that

$$\|x^k g\| \leq \pi^k \|g\|.$$

Now, let $S : l^2(Z) \to H$ be the linear operator given by

$$S((a_n)) = \sum_{n=-\infty}^{\infty} a_n (f_n - e_n), \quad (a_n) \in l^2(Z).$$

c) Assuming that the order of summation can be reversed, prove that

$$\|S\| \leq e^{M\pi} - 1.$$

Let the linear operator $K : H \to H$ be given by

$$Kf = S((f, e_n)) = \sum_{n=-\infty}^{\infty} (f, e_n)(f_n - e_n), \quad f \in H.$$

d) Show that $\|K\| < 1$ and deduce that $T = I + K \in B(H)$ has an inverse $T^{-1} \in B(H)$. Calculate Te_n and infer that

$$span\{f_n\}^{\perp} = \{0\}.$$

e) Define $g_n = (T^{-1})^* e_n$, $n \in Z$ and show that $(f_m, g_n) = \delta_{mn}$ (Kroenecker delta).

f) Finally, use the decomposition

$$T^{-1}f = \sum_{n=-\infty}^{\infty} (T^{-1}f, e_n)e_n$$

to show that

$$f = \frac{1}{\sqrt{2\pi}} \sum_{n=-\infty}^{\infty} (f, g_n)e^{i\lambda_n x}$$

for $f \in H$. □

Winter 86/87

Exercise 172
Let $T : l^2 \to l^2$ be the linear operator given by

$$T(x_1, x_2, ..., x_{2n-1}, x_{2n}, ...) = (x_2, x_1, \frac{1}{2}x_4, \frac{1}{2}x_3, ..., \frac{1}{n}x_{2n}, \frac{1}{n}x_{2n-1}, ...)$$

1) Find $\|T\|$.
2) Find T^*.
3) Find the spectrum and resolvent set for T, and determine a set of basis vectors for the eigenspace associated to $\lambda \in \sigma_p(T)$.
4) Prove that T is compact. □

Exercise 173
Let $F = \{f \in C^2([0;1]) \mid f(0) = f(1) = 0\} \subset L^2([0;1])$.
a) Show that $\|f'\|^2 \le \|f\|\|f''\|$ for $f \in F$.
b) Let $f \in F$. Show that $|f(x)| \le \|f'\|\sqrt{x}$ for $0 \le x \le 1$, and deduce that

$$\|f\| \le \frac{1}{\sqrt{2}}\|f'\|.$$

c) Show that for $f \in C^2([0;1])$ with $f(0) = f(1)$ we have

$$\|f'\| \le \frac{1}{\sqrt{2}}\|f''\|.$$

d) Show by a counterexample that the result from question (c) is not valid for general $f \in C^2([0;1])$.   ▯

Exercise 174

Let $H = L^2([0;1])$, and consider the operator K with domain $D(K) = C([0;1])$ given by

$$Kf(x) = x \int_0^x f(t)dt + \int_x^1 tf(t)dt, \quad f \in D(K).$$

a) Show that $K : D(K) \to C^2([0;1])$, and that

$$(Kf)'(0) = 0$$

and

$$(Kf)'(1) = 1.$$

b) Show that K is injective and that K^{-1} has the domain

$$D(K^{-1}) = \{u \in C^2([0;1]) \mid u'(0) = 0, \quad u(1) = u'(1)\},$$

and the action $K^{-1}u = u''$.

c) Show that K is an integral operator with continuous and symmetric kernel and find this kernel.

d) Let φ and ψ denote eigenfunctions for K associated to the same eigenvalue λ. Define the function f by

$$f(x) = \psi(0)\varphi(x) - \varphi(0)\psi(x),$$

and use the existence and uniqueness theorem for ordinary differential equations to argue that $f = 0$. Next show that all eigenspaces for K are of dimension one.

e) Let $\sigma_p(K) = (\lambda_n)$ denote the sequence of eigenvalues for K. Find

$$\sum_{n=1}^{\infty} \lambda_n^2.$$

f) Let λ be a positive eigenvalue and let $\mu = \frac{1}{\sqrt{\lambda}}$. Express the associated eigenfunction with μ and find a trancedent equation for μ. Use a graph argument to show that K has at most one positive eigenvalue.

   ▯

Winter 87/88

Exercise 175

Let H denote a Hilbert space, (e_n) an orthonormal basis, and $0 < \lambda < 1$ a constant. We define a sequence (g_n) in H by

$$g_n = \sqrt{1 - \lambda^2} e_n, \quad n \in N.$$

Find $\lim_n \|f + g_n\|$ for $f \in H$.
Define the function ξ by

$$\xi(f) = 1 \quad \text{for} \quad \|f\| < 1$$

and

$$\xi(f) = 0 \quad \text{for} \quad \|f\| \geq 1.$$

Show that $\lim_n \xi(f + g_n) = 1$ if $\|f\| < \lambda$, and $\lim_n \xi(f + g_n) = 0$ if $\|f\| > \lambda$.
(The case $\|f\| = \lambda$ is omitted). ▯

Exercise 176

In l^2 we consider the operator

$$T(x_1, x_2, ..., x_n, ...) = (2x_2, \frac{3}{2}x_3, ..., \frac{n+1}{n}x_{n+1}, ...).$$

1) Find $\|T\|$.
2) Find $\sigma_p(T)$, and find the eigenspace associated to all $\lambda \in \sigma_p(T)$.
3) Determine the adjoint T^*.
4) Determine $\sigma_r(T)$.
5) Let $\lambda \notin \sigma_p(T) \cup \sigma_r(T)$. For $k \in N$ we define an operator I_k on l^2 by

$$I_k((x_1, x_2, ..., x_k, k_{k+1}...) = (0, 0, ..., 0, x_k, k_{k+1}...),$$

and we define $T_k = I_k T$. Show that there is a $k \in N$ such that

$$\|T_k\| < \lambda.$$

Use this to solve the equation

$$(T_k - \lambda I_k)x = y$$

for a given $y \in l^2$. Finally, show that the equation

$$(T - \lambda I)x = y$$

has a solution $x = (T - \lambda I)^{-1}y$ for all $y \in l^2$.

6) Find $\sigma(T)$ and $\rho(T)$ (e.g., by use of the closed graph theorem).   ▯

Exercise 177
Let $K \in B(H)$ where $H = L^2([0; 1])$, and let K be given by

$$Kf(x) = \int_{1-x}^{1} f(t)dt.$$

1) Show that K is actually bounded.
2) Show that the kernel $k(x, t)$ for K is Hermitian and calculate

$$\|k\|^2 = \int_0^1 \int_0^1 |k(x, t)|^2 dt dx.$$

3) Show that the kernel $k_2(x, t)$ for K^2 is $\min\{x, t\}$.
4) Show that an eigenfunction for K is an eigenfunction for K^2.
Now, let f denote an eigenfunction for K associated with the eigenvalue
λ. Calculate $(K^2 f)''$, justify that it belongs to H, and show that f is a
solution to the equation
$$\lambda^2 f'' + f = 0.$$

5) Find all eigenvalues and associated eigenfunctions for K.
6) Determine $\|K\|$.
    ▯

Winter88/89

Exercise 178
a) Let $1 \leq p \leq q \leq \infty$. Show that $l^p \subset l^q$.
b) Let $1 \leq r < p < 2r$ and assume that the sequence (x_n) satisfies

$$\sum_{n=1}^{\infty} n|x_n|^p < \infty.$$

Show that $(x_n) \in l^r$.   ▯

Exercise 179
Let $p > 1$, and let $f(x, t) \geq 0$ be a (measurable) function on R^2 such that

$$g(t) = \left(\int_R f(x, t)dx \right)^{p-1}$$

exists.

1) Put $q = \frac{p}{p-1}$ and show that

$$\left\| \int_R f(x, \cdot) dx \right\|_p^p \le \|g\|_q \int_R \|f(x, \cdot)\|_p dx.$$

2) Let $f(x, t)$ be a (measurable) function on R^2 such that the function

$$x \to \|f(x, \cdot)\|_p$$

belongs to $L^1(R)$. Use Question 1 to show the inequality

$$\left\| \int_R f(x, \cdot) dx \right\|_p \le \int_R \|f(x, \cdot)\|_p dx$$

first for $p > 1$, and then for $p = 1$.

3) Let $g \in L^p(R)$ and $h \in L^1(R)$. We define the *convolution* $g * h$ by

$$g * h(t) = \int_R g(t - x) h(x) dx.$$

Show that convolution with a $L^1(R)$-function is a linear and bounded mapping from $L^p(R)$ into $L^p(R)$ for any $p > 1$. ☐

Exercise 180
Let E be a Banach space and let $P \in B(E)$ satisfy $P^2 = P$.
1) Show that $P - \lambda I$ is injective for $\lambda \in C \setminus \{0, 1\}$.
2) Show that $P - \lambda I$ is surjective for $\lambda \in C \setminus \{0, 1\}$, and find $(P - \lambda I)^{-1}$.
3) Show that $\sigma(P) = \sigma_p(P) = \{0, 1\}$.
☐

Exercise 181
Let H denote the Hilbert space $L^2([0; 2\pi])$ with the subspace $F = C([0; 2\pi])$, and let K denote the integral operator on H with the kernel

$$k(x, t) = \begin{cases} \frac{i}{2} e^{(\frac{i}{2}(x-t))} & \text{if } 0 \le t < x \le 2\pi, \\ 0 & \text{if } 0 \le t = x \le 2\pi, \\ \frac{-i}{2} e^{(\frac{i}{2}(x-t))} & \text{if } 0 \le x < t \le 2\pi. \end{cases}$$

1) Show that K is a self-adjoint Hilbert-Schmidt operator.
2) Assume that F is equipped with the sup-norm. Show that $K : H \to F$ is continuous.
3) Now let S denote the restriction of K to F (considered as a subspace of H). Show that S is injective and that S^{-1} is given by

$$D(S^{-1}) = \{g \in C^1([0; 2\pi]) \mid g(0) = g(2\pi)\}$$

and

$$S^{-1}g = -ig' - \frac{1}{2}g \quad \text{for} \quad g \in D(S^{-1}).$$

4) Find all normalized eigenfunctions and associated eigenvalues for S^{-1}. Show that all eigenvalues are simple and that the set of normalized eigenfunctions is an orthonormal system in H.

5) Show that the eigenfunctions for S^{-1} are also eigenfunctions for K, and find the associated eigenvalues. Justify that all eigenfunctions for K are given this way, and write the kernel for K using the normalized eigenfunctions.

6) Let $f \in H$ be given by the Fourier expansion

$$f = \sum_{n=-\infty}^{\infty} c_n e^{inx}.$$

Expand Kf using the Fourier coefficients c_n instead of f. $\Box$

References

[1] R.A.Adams, *Sobolev Spaces*, Academic Press, 1975.

[2] S.Agmon, A.Douglis and L.Nirenberg, Estimates near the boundary for solutions of elliptic partial differential equations satisfying general boundary conditions., *Comm. Pure Appl. Math. 12 (1959)*, pp. 623-727 and *17* (1964), pp. 35-92.

[3] J.Aubin *Applied Functional Analysis* , John Wiley & Sons Inc., 1979.

[4] C.Bardos, G.Lebeau and J.Rauch, Contrôle et stabilisation dans les problèmes hyperboliques, Appendix II in J.L.Lions [22], T.1, pp. 492-537.

[5] J.Bergh and J.Löfström *Interpolation Spaces, An Introduction*, Springer-Verlag, 1976.

[6] J.Diximier, *Les algébrés d'opérateurs dans l'espace hilbertien*, Gauthier-Villars, Paris, 1957.

[7] P.Grisvard, *Elliptic problems in non-smooth domains*, Pitman, 1985.

[8] P.Grisvard, Controllabilité excate dans les polygones et polyhédres, *C.R.Acad.Sci. Paris, série I Math., Vol. 304, n. 13, 1987*, pp. 367-370.

[9] P.Grisvard, Controllabilité excate des solutions de l'équation des ondes en présence de singularité, *J. Math. Pures et Appl., 68, 1989*, pp. 215-259.

[10] G.Grubb, *Partial Differential Equations on Domains with Boundaries*, Lecture notes, Mathematical Institute, University of Copenhagen, 1992.

[11] L.F.Ho, Observabilité frontière de l'equation des ondes, *C.R.Acad.Sci. Paris, série I Math., Vol. 302 n. 12, 1986*, pp. 443-446.

[12] O.G.Jørsboe, *Funktionalanalyse*, MAT-Noter Nr. 214, Matematisk Institut, Danmarks Tekniske Højskole, 1990.

[13] V.Komornik, Contrôlabilité exacte en un temps minimal, *C.R.Acad.Sci. Paris, série I Math., Vol. 304, 1987*, pp. 223-225.

[14] V.Komornik, Exact Controllability in Short Time for the Wave Equation, *Ann. Inst. Henri Poincaré - Analyse non linéaire, Vol.6, n.2, 1989*, pp. 153-164.

[15] J.Lagnese, Control of Wave Process with Distributed Controls supported on a Subregion, *SIAM J. Control and Optimization, Vol.21, No.1, 1983*, pp. 65-85.

[16] I.Lasiecka, J.L.Lions and R.Triggiani, Non Homogeneous Boundary Value Problems Second Order Hyperbolic Operators, *J. Math. Pures et Appl., 65, 1986*, pp. 149-192.

[17] I.Lasiecka and R.Triggiani, A cosine operator approach to modeling $L_2(0,T;L_2(\Gamma))$-boundary input hyperbolic equations, *Appl. Math. and Optim., 7(1981)*, pp. 35-83.

[18] I.Lasiecka and R.Triggiani, Regularity of hyperbolic equations under $L_2(0,T;L_2(\Gamma))$ -boundary terms, *Appl. Math. and Optim., 10(1983)*, pp. 275-286.

[19] J.L.Lions and E.Magenes, *Non-Homogeneous Boundary Value Problems and Applications Vol. I & II*, Springer-Verlag, 1972.

[20] J.L.Lions, *Contrôle des systémes distribues singuliers*, Gauthier-Villars, 1983.

[21] J.L.Lions, Exact Controllability, Stabilization and Perturbations for Distributed Systems, *SIAM Review, Vol.30, No.1, March 1988*, pp. 1-68.

[22] J.L.Lions, *Contrôlabilité exacte, stabilisation et perturbations des systemes distrubés, T. 1, Contrôlabilité exacte*, Masson, Paris, 1988, *T.2, Perturbations*, Masson, Paris, 1989.

[23] M.Renardy and R.C.Rogers, *An Introduction to Partial Differential Equations*, Springer-Verlag, 1992.

[24] F.Riesz and B.Nagy, *Vorlesungen über Funktionalanalysis*, VEB Deutscher Verlag der Wissenschaften, Berlin, 1956.

[25] W.Rudin, *Functional Analysis*, McGraw-Hill 1973.

[26] W.Rudin, *Real and Complex Analysis*, McGraw-Hill 1974.

[27] D.L.Russel, Controllability and Stabilizablility Theory for Linear Partial Differential Equations. Recent Progress and open Questions, *SIAM Review Vol.20, 1978*, pp. 639-739.

[28] J.Simon, Compact Sets in the Space $L^p(0, T; B)$, *Annali di Mat. Pura ed Appl., IV, CXLVI, Bologna, 1987*, pp. 65-96.

[29] F.Treves *Basic Linear Partial Differential Equations*, Academic Press, 1975.

[30] R.Triggiani, Exact Boundary Controllability on $L_2(\Omega) \times H^{-1}(\Omega)$ of the Wave Equation with Dirichlet Boundary Control Acting on a Portion of the Boundary $\partial\Omega$, and Related Problems, *Appl. Math. and Optim., 18, 1988*, pp. 241-277.

[31] J.Wloka, *Partial Differential Equations*, Cambridge University Press, 1992.

[32] K.Yosida, *Functional Analysis*, Springer-Verlag, 1974.

Index